# Plant Analysis Procedures
Second Edition

# Plant Analysis Procedures
## Second Edition

Edited by

**Erwin E.J.M. Temminghoff**
*Wageningen University,*
*Department of Soil Quality,*
*The Netherlands*

and

**Victor J.G. Houba**

KLUWER ACADEMIC PUBLISHERS
DORDRECHT / BOSTON / LONDON

A C.I.P. Catalogue record for this book is available from the Library of Congress.

ISBN 978-90-481-6717-3
ISBN 978-1-4020-2976-9 (e-book)

---

Published by Kluwer Academic Publishers,
P.O. Box 17, 3300 AA Dordrecht, The Netherlands.

Sold and distributed in North, Central and South America
by Kluwer Academic Publishers,
101 Philip Drive, Norwell, MA 02061, U.S.A.

In all other countries, sold and distributed
by Kluwer Academic Publishers,
P.O. Box 322, 3300 AH Dordrecht, The Netherlands.

*Printed on acid-free paper*

# PREFACE

This syllabus is a revised and extended edition of the former "Plant Analysis Manual' by I. Walinga, J.J. van der Lee, V.J.G. Houba, W. van Vark & I. Novozamsky (ISBN 0-7923-3182-6) originally meant for use within the M.Sc. course in Soil Science and Water Management and the International Postgraduate Course on Soil and Plant Analysis and Data Handling.
With this second edition of the Manual, new digestion techniques were included like microwave. Also new measure techniques were included like ICP-OES and ICP-MS. It has been reorganized and put into a new handsome format. We would like to thank Mr. Peter Nobels for the optimisation information regarding the ICP-OES and Mrs. Gerdine Gaikhorst regarding the ICP-MS. The Manual is arranged according to the methods of digestion and extraction, and all corresponding determination procedures are placed in the same chapter. Thus, each determination is tailored to the conditions of the digest/extract one is working with. For almost all elements different measure techniques are available. The elements are in alphabetic order of symbols. We hope this Manual is very useful in your laboratory during analysis of all kind of plant materials.

The authors accept no responsibility whatsoever for any harm caused by the application (right or wrong) of these procedures.

Any questions, comments, suggestions for improvements should be addressed to:

Dr. Erwin J.M. Temminghoff
Wageningen University
Department of Soil Quality
P.O. Box 8005
6700 EC Wageningen
The Netherlands

e-mail: erwin.temminghoff@wur.nl
tel: +31.(0)317.482357
fax: +31.(0)317.483766

# CONTENTS

# 1. INTRODUCTION

## 1.1 ORGANISATION OF THIS MANUAL

This manual is intended for the practising chemist who has to do a job in analysing plant material. Therefore, the present manual only contains ready-to-hand procedures without any comment. The procedures described are only for inorganic components, which frequently occur in the plant. Most procedures are designed to give a total content value of the element under consideration, regardless of the chemical structure in which it occurs in the plant.

Sampling and conservation of plant material are beyond the scope of this manual; some general remarks are, however, compiled in section 1.3. It is anyhow assumed that the material, which is handed in for analysis, is representative for the purpose of the principal.

In the field of plant analysis there is a confusing variety of methods and procedures, both for digestions and determinations; moreover, in many cases the digestion and the subsequent determination are interrelated. For example, a separate digestion using larger quantities of plant material is sometimes needed for trace elements in order to obtain measurable concentrations. After ample discussion, we have chosen for a design in which all digestion procedures are described in one chapter, all extraction procedures in one chapter and all determination procedures in one chapter. As a consequence, one has to choose a suitable digestion method in combination with the intended determination technique; this has been indicated within each individual determination procedure.

For determination of the elements, mainly spectrometric techniques are used here. Depending on the kind of element and the expected concentration level, the following methods are applied: flame atomic emission spectrometry (flame AES), flame atomic absorption spectrometry (flame AAS), inductively coupled plasma optical emission spectrometry (ICP-OES), electrothermal atomisation (graphite furnace) atomic absorption spectrometry (ETA-AAS), inductively coupled plasma mass spectrometry (ICP-MS), spectrophotometry and segmented flow analysis (SFA). Besides, potentiometry (ion selective electrodes (ISE)) and coulometry will be encountered. In many cases, more than one method is described to determine a component. This provides a reference, as well as an alternative in case of instrumental or analytical problems.

In principle, each procedure is fully written down, without reference to likewise procedures, since it is very inconvenient to be forced to riffle through the pages while performing the analysis. Often, however, one and the same determination procedure can be applied to different digests. The standard series should then be made with different acids so as to match the acid type and strength of the sample digests in question. In those cases, one procedure is described with special reference to the different media.

Many of the methods in this manual have been checked in our laboratory, which is reflected by specific remarks.

Literature references are only sparingly mentioned; in principle only if these provide specific extra information. This may be the first article on the subject in question, but it may also be a recent (review) paper as well.

In general, no warnings are given with respect to safety. It is assumed that every analyst and laboratory technician realises that all chemicals are unsafe, although to a different extent. Exceptions are only made in the case of a particularly toxic or harmful substance, e.g. KCN. For the rest it is assumed that all pipetting is done with the help of pipette fillers, that safety goggles are worn when performing digestions, that work is done in a fume hood when harmful fumes may be released, etc. There is very much literature on the subject of laboratory safety; it is strongly recommended to consult such safety manuals and to observe the safety rules given therein.

No special references are made with respect to the impact of these procedures on the environment. The user should realise, however, that the waste from a laboratory might be very detrimental for the environment, even in small amounts. Users are strongly recommended to take appropriate measures in order to minimise these effects.

## 1.2 TECHNICAL REMARKS

In applying the procedures described in this manual, one should realise that there is a number of underlying assumptions, agreements and conventions. For clarity and brevity, these are summarised below, so that there is no need to repeat them with every procedure.

- The common simple laboratory equipment and glassware (hot plates, beakers etc.) is supposed to be available. If special equipment, glassware or plastic ware is needed, it will be mentioned explicitly.
- In many cases, the classical volumetric pipette can be substituted by plunger-type pipettes or by dispensers, provided that these are regularly calibrated. This is in particular convenient for dispensing concentrated acids. For work where high precision is required, calibration of pipettes is needed.
- All reagents are assumed to be of analytical grade, unless otherwise indicated.
- It is assumed that all glassware has been cleaned thoroughly.
- When water is used, either demineralised (demi-) or ultra pure water (UPW) is meant. Ultra pure water is necessary when sensitive measuring techniques like ETA-AAS or ICP-MS will be used to measure very low concentrations.
- In the case the concentrations in plant digests are to low for flame AAS or ICP-OES one should use ETA-AAS or ICP-MS.
- Standard solutions with concentrations below 1 mg/L should be prepared in diluted acid to prevent adsorption of the standard by glass. The acid concentration should be that low, that a negligible acid concentration exists by preparation between the standard series individually and the samples.
- In the case of ICP-OES measurements scandium and/or beryllium is used as an internal standard. Differences in element concentrations in the plant digests can affect the nebuliser and/or the plasma conditions. Therefore scandium and/or beryllium will be nebulised in the plasma together with the sample. By measuring also scandium or beryllium a correction can made for all measured elements since the measured scandium or beryllium concentration should be constant.
- In the case of ICP-MS measurements scandium, germanium, rhodium, and bismuth are used as an internal standard mixture. Differences in element concentrations in the plant digests can affect the nebuliser and/or the plasma conditions. Therefore a mixture of scandium (amu 45), germanium (amu 72), rhodium (amu 103), and bismuth (amu 209) will be nebulised in the plasma together with the sample. The element within the internal standard mixtuere which has to be used for correction is the one which is close to the mass of the element of interest.
- In the author's laboratory an ICP quadropole mass spectrometer is used. By making use of a High Resolution ICP-MS less interferences are expected.
- Procedures are pointed at consumption of minimum amounts of reagents. Thus, a sentence may read "dissolve 2 g in 50 mL" when this is the normal amount for one run. You may prepare, however, 10 g in 250 mL, or any multiple, if you intend to do more runs. With automated methods, however, all reagents are given per litre, since the consumption simply depends on the number of determinations.

- The number of significant digits is an indication of the required precision. Thus, 1000 mL is far more precise than 1 litre. By applying this convention, there is no need to prescribe the required glassware, so that the user can select appropriate equipment according to his own findings. As exception to the rule, the use of volumetric flasks is explicitly prescribed for stock solutions. In practice, volumetric flasks are very convenient to handle, e.g. for mixing, so that these may be used even though their intrinsic precision is not needed.
- The atomic weights used for calculation are the most recent available, i.e., IUPAC (1987).
- SI units have been used throughout this manual. Within this system there is a preference for the use of basic units; thus, the density of water should be given in kg $m^{-3}$. For better readability, however, we have chosen for the use of g/mL if this does not cause ambiguity. We have also chosen for the consequent use of molarity in stead of normality.
- For calculation of the results, a condensed formula is given. In this way, anyone can set up his/her own way of calculating, whereas the final result can be checked by the given formula.
- The content of a compound in plant material can be expressed in (milli)moles or (milli)grams per kg. For this manual we have chosen to adhere to the generally accepted usage of mmol/kg and mg/kg for macro elements, and mg/kg or µg/kg for microelements.
- The reproducibility of determinations by the given procedures is given by a coefficient of variation (%). At low contents however a constant variation can be found therefore in many cases the reproducibility is given as a variation coefficient (%) + a constant value.
- It may be mentioned here, that any laboratory should take measures to improve and maintain the quality of its analytical results (Good Laboratory Practice, GLP). In particular, the use of certified reference and reference samples (Quality Control Charts) and an active membership of inter-laboratory trials (ringtests) are appropriate means for this purpose (Van Reeuwijk, 1998).

## REFERENCES

1. IUPAC commission on Atomic Weights and Isotopic Abundances: Atomic Weights of the Elements 1987. Pure Appl. Chem. (1988) 60: 841-854.

2. Van Reeuwijk, L.P. 1998. Guidelines for quality management in soil and plant laboratories. FAO Soils Bulletin 74. Rome 0-144. ISDN 92-5-104065-6.

# 1.3 SAMPLING AND PRETREATMENT

Although the present manual is not intended to describe sampling and pre-treatment procedures for plant material, these stages are sufficiently important to mention here the main considerations and rules.

The sample must be representative for the whole lot of plant material. This is a crucial point for the principal, since the chemical composition may vary strongly, dependent on plant part (leaf, stem, fruit), physiological age and growth conditions. Besides, it is necessary to collect as much plant material as possible in order to minimise variations due to heterogeneity.

Pre-treatment of the sample involves drying and grinding, preceded by washing if the fresh plant material is (likely to be) polluted. Soil, dust and salts from irrigation water are the usual contaminants; these may be washed out by tap water, 0.1 M hydrochloric acid or 1 % detergent solution, followed by rinsing with demi water. The internal concentration of major nutrients will not significantly be affected by this treatment, if the washing does not take more than 30 s.

The plant sample is normally dried at 70 °C in a well-ventilated drying oven till dry (often within 24 hours). The material is then finely ground, in order to obtain a homogeneous sample from which representative sub-samples can simply be taken. As a rule of thumb, the milled plant material should pass a 1-mm sieve when less than 1 gram is to be weighed out. Both drying and milling should be carried out with equipment that does not release elements for which the samples are to be analysed. One should realise, for instance, that a mill might contaminate the sample with Al, Cd, Cu, Fe, Pb and possibly other heavy metals, depending on its composition.

The dried and milled samples should be stored in a cool and dry place in tightly closed flasks or in sealed polythene bags, protected against direct sunlight. During storage, the plant material may attract moisture so that the drying procedure must be repeated just before weighing out a sample for analysis. Dry plant material can be stored for at least 10 years for the elements Al, B, Ca, Cl, Cu, Fe, K, Mg, Mn, N, Na, $NO_3^-$, P, S and Zn at laboratory conditions (Houba et al., 1995).

The analytical results are often referred to 'oven-dry' material, which means dried at 105 °C. For comparability, therefore, the moisture content should be determined by drying at 105 °C and taking the difference with the 70 °C dried sample. The drying at 105 °C should be done, however, with a separate sample, since this operation may change its chemical composition.

## REFERENCES

1. Keizer, M.G., V.J.G. Houba and Th.M. Lexmond. 1984. Sampling of soil and plant material for chemical analysis. Syllabus Department of Soil Science and Plant Nutrition, Agricultural University, Wageningen, The Netherlands (in Dutch).

2. Sonneveld, C. and P.A. van Dijk. 1982. The effectiveness of some washing procedures in the removal of contaminants from plant tissue samples of glasshouse crops. Commun.Soil Sci. Plant Anal. 13: 487-496.

3. Houba, V.J.G., I. Novozamsky and J.J. van der Lee. 1995. Influence of storage of plant samples on their chemical composition. Sci. of the Env. 176: 73-79.

# 2. DIGESTIONS

## 2.1 DIGESTION IN TUBES WITH $H_2SO_4$ - SALICYLIC ACID - $H_2O_2$ AND SELENIUM

### 1. FIELD OF APPLICATION

1.1 This digestion is in particular suited for routine work on large series of plant samples followed by automated determinations. It can be applied for the determination of total calcium (Ca), potassium (K), magnesium (Mg), manganese (Mn), nitrogen (N), sodium (Na), phosphorus (P), and zinc (Zn) in plant material.

### 2. PRINCIPLE

2.1 The larger part of organic matter is oxidised by hydrogen peroxide at relatively low temperature. Salicylic acid is added to prevent loss of nitrate. After decomposition of the excess $H_2O_2$ and evaporation of water, the digestion is completed by concentrated sulphuric acid at elevated temperature under the influence of Se as a catalyst.

### 3. APPARATUS

3.1 Aluminium heating block with holes for digestion tubes.

3.2 Metal weighing funnels with long spouts (Figure 1).

3.3 Digestion tubes, 100 mL, with narrowed neck (Figure 2).

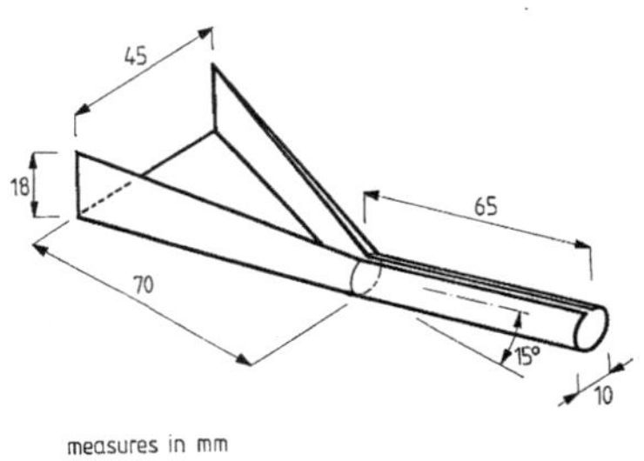


Figure 1 Weighing funnel.

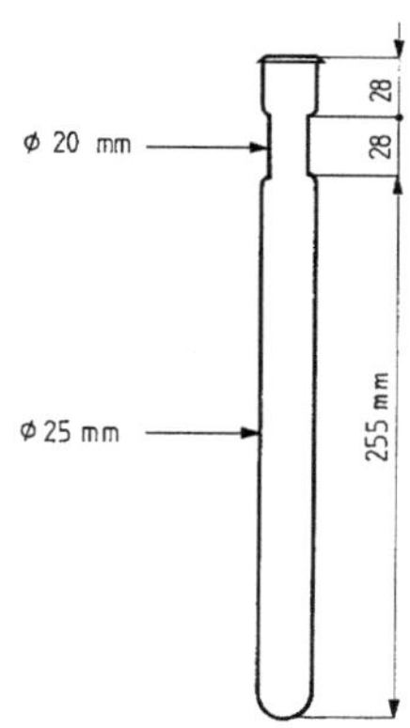


Figure 2 Digestion tube.

Remarks:

1. Dried plant material may easily stick to glass when the relative humidity of the air is low. These homemade weighing funnels (stainless steel or aluminium) do not show this effect.
2. These weighing funnels have been designed with an extra long spout, so that the plant material is released below the narrowed neck.
3. The dimensions of digestion tubes may be different from those indicated in Figure 2, as long as they fit exactly into the holes of the heating block used, have a narrowed neck and a length of at least 15 cm.
4. In the authors' laboratory, a series consists of 40 tubes: 34 samples, 2 blanks, 1 plant sample with known low concentration (in duplicate) and 1 with known high concentration (also in duplicate). These known samples serve as internal quality control.

## 4. REAGENTS

4.1 Sulphuric Acid, 96 % (w/w), 18 mol/L ($\rho$ = 1.84 g/cm$^3$).

4.2 Hydrogen Peroxide, 30 % (w/w).

4.3 Selenium, Powder.

4.4 Salicylic Acid, Powder.

4.5 Sulphuric Acid - Selenium Mixture - Dissolve 3.5 g of selenium (4.3) in 1 litre of sulphuric acid (4.1) by heating to about 300 °C, while covering the beaker with a watch glass. The originally black colour of the suspension turns via green/blue into clear light yellow. The entire process takes 3-4 hours.

4.6 Digestion mixture - Dissolve 7.2 g of salicylic acid (4.4) in 100 mL of the sulphuric acid - selenium mixture (4.5). This digestion mixture should not be stored for more than 48 h.

Remarks:

5. In the authors' routine laboratory, where this digestion is applied every day, the plant material is dried again at 70 °C just before weighing; at that moment its moisture content is only 1-2 %, so that concentrated sulphuric acid may be used for the digestion. When the drying is not repeated, the plant sample may contain up to 10 % moisture. The use of concentrated sulphuric acid then causes a raise in temperature, which results in loss of nitrate. In that case, diluted sulphuric acid should be used as described in section 2.2 (digestion in flasks).
6. Hydrogen peroxide must be of analytical quality; a lower grade may be stabilised with EDTA, phosphate, or other interfering compounds.

## 5. PROCEDURE

5.1 Weigh, to the nearest 0.001-g, approximately 0.3 g of the dried plant material sample in a metal weighing funnel and transfer the sample to a digestion tube. Take care that all the plant material comes below the narrow part of the tube. Add 2.5 mL of the digestion mixture (4.6) and swirl carefully until all the plant material is moistened. Allow standing for at least 2 h. Prepare also two blank digestions.

Place the tube in the heating block and heat at 100 °C for at least 2 h. Remove the tube from the block, allow to cool, and add successively three 1-mL aliquots of hydrogen peroxide (4.2). Mix carefully, but thoroughly after each addition. **WARNING**: the reaction is violent. Wait until the reaction with hydrogen peroxide has ceased (approximately 10 s) before adding the next portion. Place the tube in the preheated heating block at 330 °C. Wait for 1 h and if the colour of the digest is deep yellow, cool and add another 1mL of hydrogen peroxide (4.2). Place the tube again in the preheated block and heat at 330 °C. The digestion is considered complete when the digests have turned colourless or light-yellow; this usually takes about 2 h. Remove the tube from the block and cool to room temperature. Add 48.3 mL of water and mix. Allow to stand overnight. Mix again, transfer the digest to a test tube and let settle.

Remarks:

7. After moistening the sample with digestion mixture (4.6), at least 2 h are needed for the nitro-salicylic acid compounds to be formed; in the authors' laboratory this period is often taken overnight.
8. At least 2 h at 100 °C is necessary to obtain complete reduction of the nitro-salicylic acid compounds.
9. The boiling point of the digestion mixture (4.6) lies just above 330 °C; when the block temperature is adjusted at 330 °C, the temperature in the tube will be approximately 300 °C.
10. The volume of digest is remarkably constant: 2.1 ± 0.1 mL for the samples, and 2.5 mL for the blanks. For that reason, it is appropriate to add a fixed volume of 48.3 mL water to the digest by means of a dispenser. According to our experience, any volume marks on tubes lead to fairly large errors, since the tubes are wide and the marks are often inaccurately placed.
11. When Zn is to be determined, no rubber stoppers should be used.
12. $SiO_2$ will dissolve gradually and may then interfere in the determinations. Thus, the determinations should be done soon after digestion, otherwise the digest has to be filtered soon after digestion.
13. A precipitate of $CaSO_4$ may be formed when cooling after completing the digestion; it will dissolve in 18-20 hours after the addition of water. Therefore, Ca can be measured only after this period.
14. Near the end of the digestion, the liquid becomes yellow due to the last portions of organic matter, or to Fe. If the yellow colour persists after extra addition(s) of $H_2O_2$, then the presence of Fe should be assumed and the digestion can be considered as completed.
15. The filtered digest can be stored for a maximum of two weeks before analysis.
16. The calibration solutions for analysis have to be prepared in the same final medium as the samples in order to get a matrix, which is the same as in the samples. The final medium is 0.8 M $H_2SO_4$.

## 6. REFERENCES

6.1 Novozamsky, I., V.J.G. Houba, R. van Eck and W. van Vark. 1983. A novel digestion technique for multi-element plant analysis. Commun. Soil Sci. Plant Anal. 14: 239-249.

# 2.2 DIGESTION IN FLASKS WITH $H_2SO_4$ - SALICYLIC ACID - $H_2O_2$

## 1. FIELD OF APPLICATION

1.1 This digestion is intended for incidental, small series of plant samples. It can be applied for the determination of total calcium (Ca), potassium (K), magnesium (Mg), manganese (Mn), nitrogen (N), sodium (Na), phosphorus (P), and zinc (Zn) in plant material.

## 2. PRINCIPLE

2.1 With very strong sulphuric acid, the plant material is dehydrated and most of the organic matter oxidised at rather high temperature. The digestion is completed by hydrogen peroxide at elevated temperature.

Remark:
1. Salicylic acid is added to prevent loss of nitrate.

## 3. APPARATUS

3.1 Hot plate, able to reach 300 °C.

3.2 Metal weighing funnel (Figure 1).

Remark:
2. Dried plant material may easily stick to glass when the relative humidity of the air is low. These homemade weighing funnels (stainless steel or aluminium) do not show this effect.

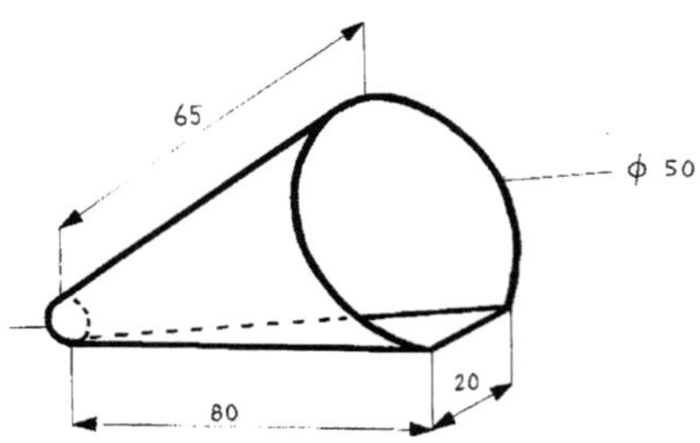


Figure 1 Weighing funnel (measures in mm).

# 4. REAGENTS

4.1 Sulphuric Acid, 96 % (w/w), 18 mol/L ($\rho$ = 1.84 g/cm$^3$).

4.2 Hydrogen Peroxide, 30 % (w/w).

4.3 Salicylic Acid, Powder.

4.4 Digestion Mixture - Put 18 mL water in a 250-mL erlenmeyer flask. While cooling, add in small portions 100 mL of sulphuric acid (4.1) (**CAUTION**). Then dissolve 6 g of salicylic acid (4.3) with the aid of a magnetic stirrer.

4.5 Carborundum Beads

Remarks:

3. Hydrogen peroxide must be of analytical quality; a lower grade may be stabilised with EDTA, phosphate, or other interfering compounds.
4. Since the dried sample still contains about 10 % (w/w) water, the sulphuric acid is somewhat diluted to prevent the temperature from raising too high during addition of the digestion mixture which otherwise would result in loss of nitrate (see section 2.1, remark 5).

# 5. PROCEDURE

5.1 Weigh, to the nearest 0.001 g, approximately 0.3 g of the dried plant material sample in a metal weighing funnel and transfer the sample to a 50-mL volumetric flask. Take care that all the plant material comes below the neck of the flask. Add 3.3 mL of the digestion mixture (4.4) and 4 carborundum beads (4.5) and swirl carefully until all the plant material is moistened. Allow to stand overnight. Prepare also two blank digestions.

Heat the flask on a hot plate at 180 °C for about 1 h. Remove the flask from the plate, let cool down, and add 5 drops of hydrogen peroxide (4.2). Place the flask on the hot plate and increase the temperature to about 280 °C. Heat 5-10 min until the water has evaporated (appearance of white vapours). Remove the flask from the plate, let cool down, add 5 drops of hydrogen peroxide (4.2), and heat again for 5-10 min to appearance of white vapours. Repeat this treatment until the digest has turned colourless.

Remove the flask from the plate and cool to room temperature. Add about 10 mL water and mix; swirl until most of the precipitate has dissolved. Make up to the mark with water, mix well and filter over coarse filter paper.

Remarks:

5. After moistening the sample with the digestion mixture (4.4), at least 2 h is needed to form the nitro-salicylic acid compounds. A longer period (overnight), however, will prevent foaming later on.
6. When heating at 180 °C, the sample will turn black and foam may be formed. If this reaches the neck of the flask, 1 or 2 drops of hydrogen peroxide (4.2) should be added.
7. A precipitate may be formed when cooling after completing the digestion. The addition of 10 mL water then produces enough heat to dissolve the precipitate rapidly.
8. The digest is filtered to remove any $SiO_2$ that will otherwise dissolve gradually and then interfere in the determinations.

9. A precipitate of $CaSO_4$ may be formed when cooling after completing the digestion; it will dissolve in 18-20 hours after the addition of water. Therefore, Ca can be measured only after this period.
10. Do not use the volumetric flasks for other purposes.
11. Near the end of the digestion, the liquid becomes yellow due to the last portions of organic matter, or to Fe. If the yellow colour persists after extra addition(s) of $H_2O_2$, then the presence of Fe should be assumed and the digestion can be considered as completed.
12. The calibration solutions for analysis have to be prepared in the same final medium as the samples in order to get a matrix, which is the same as in the samples. The final medium is 0.8 M $H_2SO_4$.

# 2.3 MICROWAVE DIGESTION WITH $HNO_3$ - $H_2O_2$ - HF

## 1. FIELD OF APPLICATION

1.1 This digestion was developed for elemental analysis of plant tissue by means of spectrometric methods (Flame-AES, Flame-AAS, ETA-AAS, ICP-OES, or ICP-MS) only. It can be applied for the determination of total aluminium (Al), arsenic (As), calcium (Ca), cadmium (Cd), cobalt (Co), chromium (Cr), copper (Cu), iron (Fe), potassium (K), magnesium (Mg), manganese (Mn), sodium (Na), nickel (Ni), phosphorus (P), lead (Pb), sulphur (S), antimony (Sb), tin (Sn), vanadium (V) and zinc (Zn) in plant material.

## 2. PRINCIPLE

2.1 Hydrofluoric acid is allowed to react first and evaporate prior to the addition of hydrogen peroxide and nitric acid with subsequent microwave heating in a closed system. Complete dissolution is achieved with a high degree of mineralisation. The fluoride is added in order to solubilise the plant tissue by destroying its silicate skeleton. Nitric acid oxidises most of the organic matter whereas hydrogen peroxide is needed to break down fatty components.

Remark:

1. The use of hydrofluoric acid, nitric acid and hydrogen peroxide in one mixture followed by the microwave heating does not lead to complete dissolution in many cases.
2. Treatment with HF is necessary to release any compounds absorbed to or included in the silicates. As a consequence Si and B cannot be determined.
3. The oxidation is incomplete, and therefore the digest can only be successfully analyzed by techniques such as AAS, AES, ICP-OES or ICP-MS.

## 3. APPARATUS

3.1 Metal weighing funnel (Figure 1).

3.2 Conventional hot plate.

3.3 Aluminium heating block, 50 mm thick, with cylindrical holes in which the digestion vessels fit.

3.4 Microwave with lined digestion vessels (PTFE inner vessel with a polyetherimide outer vessel). In the author's lab a CEM MDS 2100 microwave (1300 W output) is used.

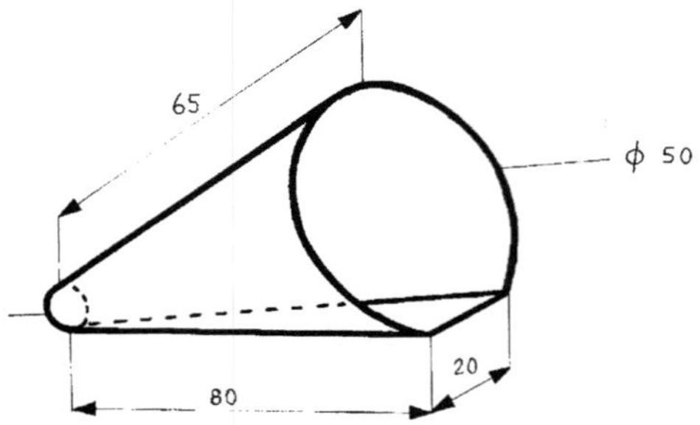


Figure 1 Weighing funnel (measures in mm).

Remark:

4. Dried plant material may easily stick to glass when the relative humidity of the air is low. The homemade weighing funnels (stainless steel or aluminium) do not show this effect.

# 4. REAGENTS

4.1 Hydrofluoric acid, 40 % (w/w), 22.6 mol/L ($\rho = 1.13\ g/cm^3$).

4.2 Nitric Acid, 65 % (w/w), 14.4 mol/L ($\rho = 1.40\ g/cm^3$).

4.3 Hydrogen Peroxide, 30 % (w/w).

4.4 Ammonia Solution 4 mol/L - Dilute 75 mL of concentrated aqueous ammonia (25 %) with water to 250 mL. See remark 6.

# 5. PROCEDURE

5.1 Weigh, to the nearest 0.001 g, approximately 0.3-0.4 g of the dried plant material sample in a metal weighing funnel and transfer the sample to the PTFE digestion vessel. Prepare also two blank digestions.
Add 4.0 mL of the hydrofluoric acid (4.1), 4.0 mL nitric acid (4.2) and mix well. Put the vessels into the holes of an aluminium block and let stand overnight at room temperature. Evaporate to almost dry using a heating temperature of 120 °C.

Add 4.0 mL of concentrated nitric acid (4.2) and three times 1.0 mL of hydrogen peroxide (4.3). Swirl carefully to moisten all the plant material and put the digestion inner vessel into the polyetherimide outer vessel. Close the vessels (hand tight) and place them in the digestor in the right sequence. Connect the pressure and temperature sensor tube to the pilot vessel. Check everything to be sure that the carrousel can turn free. Switch on the microwave and start the digestion program (Table 1). Allow the vessels to cool down for 45 min after the program is finished or until the pressure is less than 60 psi. Open the vessels in a fume hood.

Table 1 Operating parameters of the CEM microwave.

| Stage | 1 | 2 | 3 | 4 |
|---|---|---|---|---|
| Power (%) | 25 | 37 | 50 | 75 |
| Maximum pressure (psi) | 20 | 75 | 150 | 175 |
| Run time (min) | 1 | 5 | 5 | 10 |
| Maximum temperature (°C) | 100 | 125 | 150 | 175 |
| Fan speed (%) | 50 | 50 | 50 | 50 |

Transfer quantitatively – with the help of a polythene funnel – to a 50-mL polythene volumetric flask. Make up to the mark and mix. Then filter over fine paper into a polythene bottle.

Remarks:

5. The first vessel is the pilot vessel and should contain a standard plant sample of approximately 0.500 g. This is to be sure that the pressure in the other vessels is a slightly less.
6. The evaporation of HF is slow in the beginning, but very quick in the end. Therefore, this process should be observed frequently. As soon as the liquid has evaporated to dryness, the PTFE digestion vessel should be lifted from the block in order to prevent any charring.
7. Hydrofluoric acid is a treacherous skin poison. Work in a good fume hood and wear rubber gloves, safety goggles and protective cloths. If any HF comes into contact with the skin, wash immediately and thoroughly with water and thereafter dab with 4 M ammonia (4.4) or calcium gluconate gel.
8. Care should be taken that no solution drips out when the lid of the inner vessel is opened.
9. The calibration solutions for analysis have to be prepared in the same final medium as the samples in order to get a matrix, which is the same as in the samples. Since the first additions of HF and $HNO_3$ are evaporated the final medium is 1.4 M $HNO_3$.

## 6. REFERENCES

6.1 Novozamsky, I., R. van Eck, V.J.G. Houba and J.J. van der Lee. 1996. Solubilization of plant tissue with nitric acid - hydrofluoric acid - hydrogen peroxide in a closed system microwave digestor. Commun. Soil Sci. Plant Anal. 27: 867-875.

# 2.4 DIGESTION WITH $HNO_3$ - $H_2O_2$ - HF

## 1. FIELD OF APPLICATION

1.1 This digestion was developed for elemental analysis of plant tissue by means of spectrometric methods (Flame-AES, Flame-AAS, ETA-AAS, ICP-OES, or ICP-MS) only. It can be applied for the determination of total aluminium (Al), arsenic (As), calcium (Ca), cadmium (Cd), cobalt (Co), chromium (Cr), copper (Cu), iron (Fe), potassium (K), magnesium (Mg), manganese (Mn), sodium (Na), nickel (Ni), phosphorus (P), lead (Pb), sulphur (S), antimony (Sb), tin (Sn), vanadium (V) and zinc (Zn) in plant material.

## 2. PRINCIPLE

2.1 Boiling nitric acid oxidises most of the organic matter. Hydrogen peroxide is needed to break down fatty components. The fluoride is added in order to solubilise the plant tissue by destroying its silicate skeleton.

<u>Remark:</u>

1. The oxidation is incomplete, and therefore the digest can only be successfully analyzed by techniques such as AAS, AES, ICP-OES or ICP-MS.
2. Treatment with HF is necessary to release any compounds absorbed to or included in the silicates. As a consequence Si and B cannot be determined.

## 3. APPARATUS

3.1 Metal weighing funnel (Figure 1).

3.2 Conventional hot plate.

3.3 Aluminium heating block, 50 mm thick, with cylindrical holes in which the digestion vessels fit.

3.4 PTFE digestion vessels, homemade, fitting the heating block holes (Figure 2).

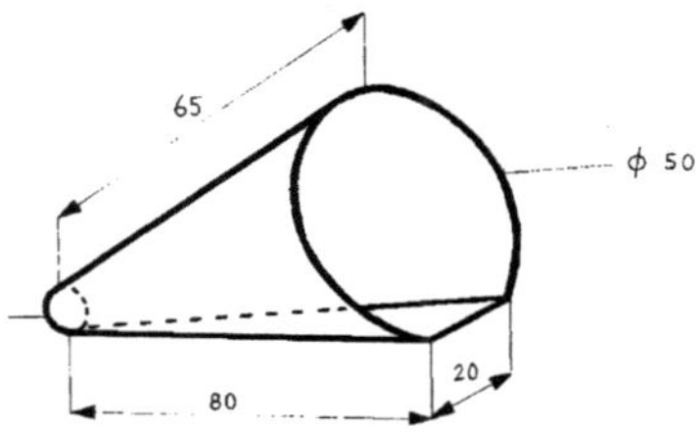


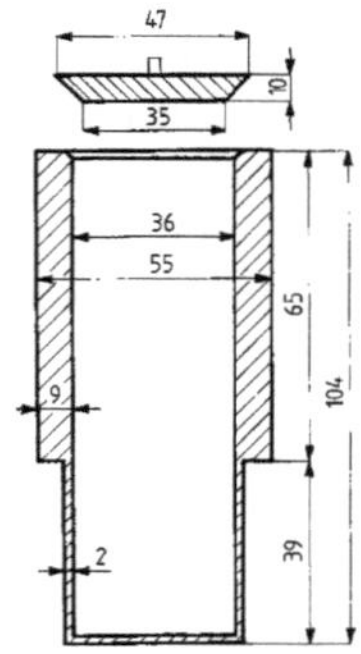


Figure 1 Weighing funnel (measures in mm).

Figure 2 Digestion vessel (measures in mm).

Remarks:

2. Dried plant material may easily stick to glass when the relative humidity of the air is low. The homemade weighing funnels (stainless steel or aluminium) do not show this effect.
3. The part of the PTFE digestion vessel that sinks in the heating block has a thin wall for better heat transfer, whereas the upper part has a thick wall for better cooling. For the same purpose, the upper part must be high enough.

## 4. REAGENTS

4.1 Nitric Acid, 65 % (w/w), 14.4 mol/L ($\rho = 1.40$ g/cm$^3$).

4.2 Hydrogen Peroxide, 30 % (w/w).

4.3 Hydrogen Fluoride, 40 % (w/w), 22.6 mol/L ($\rho = 1.13$ g/cm$^3$).

4.4 Nitric Acid Solution 2.0 mol/L - Add 138 mL of nitric acid (4.1) to about 400 mL water and make up to 1 litre.

4.5 Ammonia Solution 4 mol/L - Dilute 75 mL of concentrated aqueous ammonia (25 %) with water to 250 mL. See remark 7.

## 5. PROCEDURE

5.1 Weigh, to the nearest 0.001 g, approximately 2 g of the dried plant material sample in a metal weighing funnel and transfer the sample to the PTFE digestion vessel. Add 15 mL of concentrated nitric acid (4.1) and 5.0 mL of hydrogen fluoride (4.3). Swirl carefully to moisten all the plant material. Cover the digestion vessel with its lid and place the digestion vessel in a hole of the heating block. Allow the samples to stand overnight. Boil for 4 hours at 120 °C (see remark 4). Prepare also a blank digestion.
Remove the lid and add 1 mL of hydrogen peroxide (4.2) to the still hot solution. Wait until the reaction has ceased (about 10 seconds) and keep on heating. Repeat the

addition of hydrogen peroxide twice, and swirl. Then add 10.0 mL of concentrated nitric acid (4.1), replace the lid and continue heating for 4 hours. Thereafter, remove the lid, set the heating control at 140 °C and allow to evaporate until dry (**CAREFUL** see remark 5).

Take up the residue in 20 mL of diluted nitric acid (4.4). Lower the temperature of the block and heat for 5-10 min, taking care that the solution does not start boiling. Allow cooling, then transfer quantitatively – with the help of a polythene funnel – into a 100-mL polythene volumetric flask. Make up to the mark. Then filter over fine paper into a polythene bottle.

Remarks:

4. The samples must be left with the acid mixture at room temperature for several hours (to prevent foaming later on), and then be heated for 4 h at 120 °C. This can best be effected during the night by using a timer. Since it takes about 0.5 h to reach the temperature of 110 ºC, the timer should be set at such a time that the heating starts just 4.5 h before continuing the next morning with the rest of the procedure.
5. The evaporation of HF is slow in the beginning, but very quick in the end. Therefore, this process should be observed frequently. As soon as the liquid has evaporated to dryness, the PTFE digestion vessel should be lifted from the block in order to prevent any charring.
6. In this procedure, the lid is essential to prevent premature evaporation of nitric acid.
7. Hydrofluoric acid is a treacherous skin poison. Work in a good fume hood and wear rubber gloves, safety goggles and protective cloths. If any HF comes into contact with the skin, wash immediately and thoroughly with water and thereafter dab with 4 M ammonia (4.5) or calcium gluconate gel.
8. The calibration solutions for determination have to be prepared in the same final medium solution as the samples in order to get a matrix, which is the same as in the samples. Since the first additions of HF and $HNO_3$ are evaporated the final medium is 0.4 M $HNO_3$.

## 6. REFERENCES

6.1 Novozamsky, I., V.J.G. Houba and J.J. van der Lee. 1993. A convenient method of digestion for elemental analysis of soil and plant material. Commun. Soil Sci. Plant Anal. 24: 2595-2605.

# 2.5 DIGESTION WITH $HNO_3$

## 1. FIELD OF APPLICATION

1.1 This digestion was especially developed for the ICP-OES determination of total sulphur (S).

## 2. PRINCIPLE

2.1 Boiling nitric acid oxidises most of the organic matter.

Remark:

1. This digestion is incomplete, but the extremely high temperature of the ICP causes all undigested compounds to atomise.

## 3. APPARATUS

3.1 Metal weighing funnel (Figure 1).

3.2 Conventional hot plate.

3.3 Aluminium heating block, 50 mm thick, with cylindrical holes in which the digestion vessels fit.

3.4 PTFE tubes, homemade, precisely fitting the heating block holes (Figure 2).

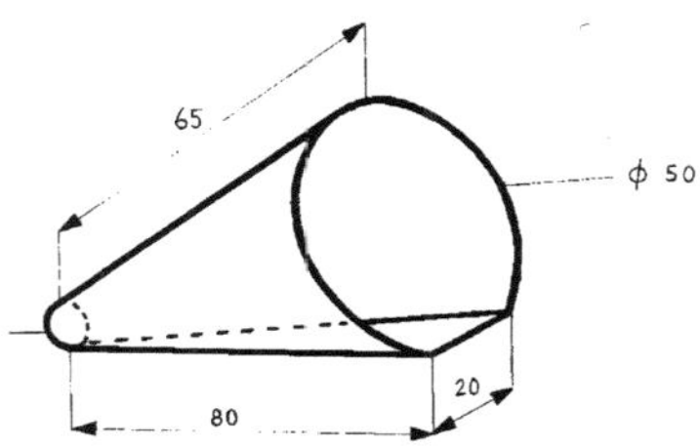


Figure 1 Weighing funnel. (measures in mm).

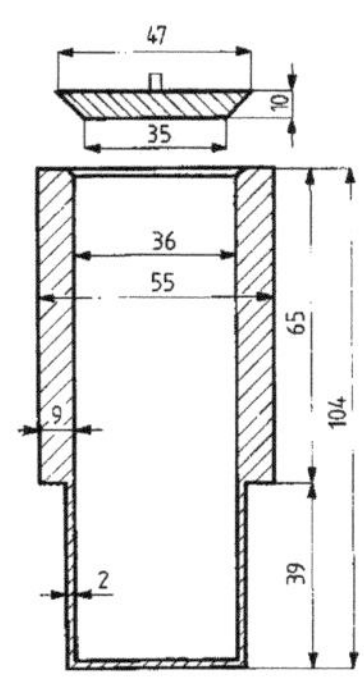


Figure 2 Digestion tube. (measures in mm).

Remarks:

2. Other heating b ocks and tubes may be used, provided that the tubes just fit the holes for efficient heat transfer, and rise high enough above the block for cooling.
3. Dried plant material may easily stick to glass when the relative humidity of the air is low. These homemade weighing funnels (stainless steel or aluminium) do not show this effect.

## 4. REAGENTS

4.1 Nitric Acid, 65 % (w/w), 14.4 mol/L ($\rho$ = 1.40 g/cm$^3$).

4.2 Nitric Acid Solution 4 mol/L - Add 277 mL of nitric acid (4.1) to about 400 mL water and make up to 1 litre.

## 5. PROCEDURE

5.1 Weigh, to the nearest 0.001 g, approximately 1 g of the dried plant material sample in a metal weighing funnel and transfer the sample to the PTFE tube provided. Add 10 mL of concentrated nitric acid (4.1) and swirl the tube carefully to moisten all the plant material. Cover the tube with its lid and allow to stand overnight. Prepare also a blank digestion.

Place the tube in a hole of the heating block and boil for 4 h at 120 °C. Remove the lid and allow the liquid to evaporate until about 1 mL remains. Take up the residue in 5 mL of nitric acid (4.2), boil for 10 min, let cool down, transfer to a 100-mL volumetric flask and make up to the mark. Filter over fine paper.

Remark:

4. The calibration solutions for analysis have to be prepared in the same digestion solution as the samples in order to get a matrix, which is the same as in the samples. The final medium is 0.2 M $HNO_3$.

## 6. REFERENCES

6.1 Novozamsky, I., R. van Eck, J.J. van der Lee, V.J.G. Houba and E. Temminghoff. 1986. Determination of total sulphur and extractable sulphate in plant materials by inductively coupled plasma atomic emission spectrometry. Commun. Soil Sci. Plant Anal. 17: 1147-1157.

# 2.6 DIGESTION WITH $HNO_3$ - $HCLO_4$ - $H_2SO_4$

## 1. FIELD OF APPLICATION

1.1 In this digestion a fairly large amount of plant material is weighed out, which allows the determination of the trace elements. It can be applied for the determination of total cadmium (Cd), copper (Cu), iron (Fe), manganese (Mn), and zinc (Zn).

## 2. PRINCIPLE

2.1 Easily oxidisable organic matter is attacked by $HNO_3$ at relatively low temperature. After the excess $HNO_3$ is distilled off, HClO4 oxidises the remaining organic compounds at elevated temperature.

Remarks:

1. Sulphuric acid is present to "dilute" the perchloric acid in the final stage so as to prevent the risk of explosions.
2. Special safety precautions are necessary when using perchloric acid
3. This digestion method is also known as the "Schaumlöffel" method (Schaumlöffel, 1960).

## 3. APPARATUS

3.1 Heating mantles or blocks for 100-mL flasks.

3.2 Metal weighing funnel (Figure 1).

3.3 Digestion apparatus adapted from Chat (Chat, 1966) (Figure 2).

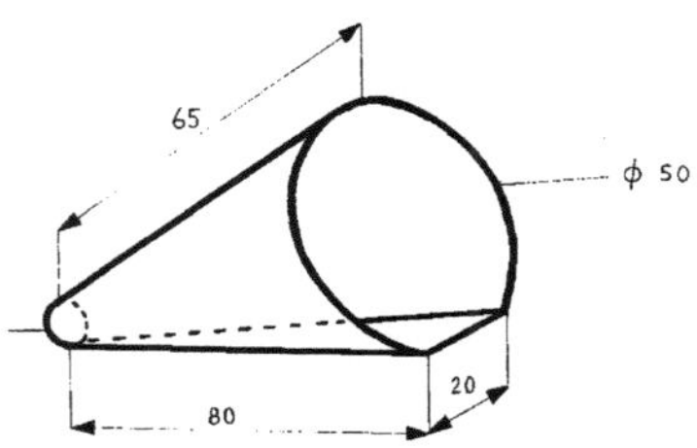


Figure 1 Weighing funnel (measures in mm).

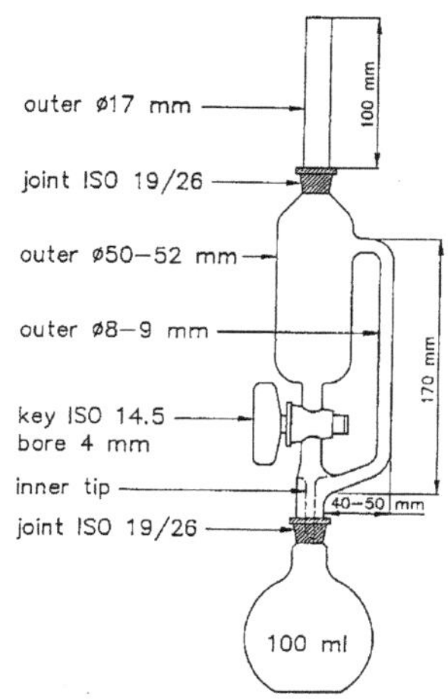


Figure 2 Digestion apparatus (measures in mm).

Remarks:

4. Dried plant material may easily stick to glass when the relative humidity of the air is low. These homemade weighing funnels (stainless steel or aluminium) do not show this effect.
5. The Chat digestion apparatus as shown in Figure 2 is commercially available. The dimensions of similar glassware from other suppliers may be slightly different, without affecting the digestion process. The stopcock is not necessary, so that a similar condenser (either purchased commercially or homemade) can be used as well.

# 4. REAGENTS

4.1 Nitric Acid, 65 % (w/w), 14.4 mol/L ($\rho = 1.40$ g/cm$^3$).

4.2 Perchloric Acid, 70 % (w/w), 11.6 mol/L ($\rho = 1.67$ g/cm$^3$).

4.3 Sulphuric Acid, 96 % (w/w), 18 mol/L ($\rho = 1.84$ g/cm$^3$).

4.4 Digestion Mixture - Mix 400 mL of nitric acid (4.1) with 40 mL of perchloric acid (4.2) and add 10 mL of sulphuric acid (4.3).

4.5 Sodium Nitrite Solution 5 g/L - Dissolve 0.5 g of $NaNO_2$ in 100 mL of water.

4.6 Carborundum Beads.

# 5. PROCEDURE

5.1 Weigh, to the nearest 0.001 g, approximately 2 g of the dried plant material sample in a metal weighing funnel and transfer the sample to the flat-bottomed 100-mL flask of the digestion apparatus. Add 20 mL of the digestion mixture (4.4) and 4 carborundum beads (4.6). Wet the plant material by swirling the flask. Moisten the lower cone of the condenser with water and place it on the flask. Moisten the cone of the upper glass tube with water and place it on top of the condenser. The stopcock should be closed. Fix the whole digestion apparatus upright and allow to stand overnight at room temperature. Prepare also two blank digestions.

Heat moderately (about 170 °C) for at least 40 min until most of the nitric acid has distilled off. Then raise the temperature stepwise, so that the remainder of nitric acid, together with some water, distils over. During this process the contents of the flask turns black. When the distillation stops, the temperature will increase until the azeotropic boiling point of perchloric acid (205 °C) is reached. The now concentrated perchloric acid attacks remaining organic material with a violent oxidation reaction, in which dense white fumes are developed. Adjust the heat supply in such a way that these perchloric acid vapours condense halfway the side arm. When the digest has turned colourless, or slightly coloured, continue digestion for 1 h more. Then cool a little; add about 20 mL water and 2 mL of sodium nitrite solution (4.5). Boil for 10 min, cool, discard the contents of the condenser, and rinse the side arm of the condenser, thereby collecting all liquid in the flat-bottomed digestion flask. Transfer the contents of this flask to a 100-mL volumetric flask; wash and collect these washings in the same volumetric flask and make up to the mark. Mix and filter over coarse paper into a 100-mL polythene bottle.

Remarks:

6. The plant material has to stand overnight with the digestion mixture to prevent excessive foaming.
7. The first stage of the digestion should take at least 40 min to allow the nitric acid to destroy all the easily oxidisable material. This prevents any explosions due to too rapid reaction by perchloric acid.
8. The hot plates in the authors' laboratory are set at position 4-5 for 170 °C and at 5.5-6.5 for 205 °C.
9. Sodium nitrite reduces insoluble higher oxides of manganese, which might have been formed.
10. Boiling is required to dissolve salts after the digestion.
11. The cones and stopcock must never be greased, but should be moistened with water.
12. The calibration solutions for analysis have to be prepared in the same digestion solution as the samples in order to get a matrix, which is the same as in the samples. Since the additions of $HClO_4$ and $HNO_3$ are evaporated the final medium is 0.08 M $H_2SO_4$.

## 6. REFERENCES

6.1 Schaumlöffel, E. 1960. Über die colorimetrische Bestimmung der Mikronährstoffe Kupfer, Zink, Kobalt, Mangan, Eisen und Molybdän aus einer Aschenlösung durch fraktionierte Extraktion. Landwirtschaftliche Forschung 13: 278-286 (in German).

6.2 Chat, G. 1966. Nouvelle méthode de minéralisation des végétaux pour les analyses chimiques. Académie d'Agriculture de France, procès-verbal de la séance du 9 Novembre 1966: 1087-1093 (in French).

# 2.7 DIGESTION BY DRY-ASHING FOLLOWED BY TREATMENT WITH HF

## 1. FIELD OF APPLICATION

1.1 This digestion was developed for elemental analysis of plant tissue by means of spectrometric methods (Flame-AES, Flame-AAS, ETA-AAS, ICP-OES, or ICP-MS) only. It can be applied for the determination of total aluminium (Al), arsenic (As), calcium (Ca), cadmium (Cd), cobalt (Co), chromium (Cr), copper (Cu), iron (Fe), potassium (K), magnesium (Mg), manganese (Mn), sodium (Na), nickel (Ni), phosphorus (P), lead (Pb), sulphur (S), antimony (Sb), tin (Sn), vanadium (V) and zinc (Zn) in plant material.

## 2. PRINCIPLE

2.1 The organic matrix of the plant material is destroyed by controlled heating. Silicates are removed as volatile fluorides.

Remarks:

1. The required high temperature causes various nitrogen compounds to volatilise.
2. Treatment with HF is necessary to release any compounds adsorbed to or included in silicates. As a consequence, Si and B cannot be determined.

## 3. APPARATUS

3.1 Platinum or sintered aluminium oxide ("Alsint") crucibles.

3.2 Muffle furnace.

3.3 Hot plate.

3.4 Stirring rods.

## 4. REAGENTS

4.1 Hydrofluoric Acid, 40 % (w/w), 22.6 mol/L ($\rho = 1.13\ g/cm^3$).

4.2 Hydrochloric Acid, 36 % (w/w), 12.0 mol/L ($\rho = 1.19\ g/cm^3$).

4.3 Ammonia Solution 4 mol/L - Dilute 75 mL of concentrated aqueous ammonia (25 %) with water to 250 mL. See remark 5.

# 5. PROCEDURE

5.1 Weigh, to the nearest 0.001 g, approximately 2 g of the dried plant material sample in a platinum or Alsint crucible. Place it in a muffle furnace at room temperature. Prepare also a blank digestion.

Raise the temperature gradually in 2 h to 450 °C, and maintain this temperature during 2 h. Take the crucible out of the furnace and allow it to cool.

Add 2-3 mL of water and then carefully dropwise 1 mL of hydrochloric acid (4.2). Wait until the production of gas bubbles has stopped. Heat gently for some minutes on a hot plate (max. 80 °C) until the appearance of fumes. Cool, add 2-3 mL of water and filter over fine paper into a 100-mL volumetric flask. Wash with lukewarm water till acid-free (< 50 mL total).

Put the filter paper and its contents into the same platinum or Alsint crucible and warm it gently in an oven until dry. Then place the crucible in the cold muffle furnace and raise the temperature to 550 °C. Maintain this temperature during 0.5 h. Take the crucible out of the furnace and allow to cool.

Add 5 mL of hydrofluoric acid (4.1). Heat the crucible on a hot plate until dry, taking care that the liquid does not boil (temperature about 100 °C). Allow to cool. Add 1 mL of hydrochloric acid (4.2), heat until appearance of fumes, cool, add 2-3 mL of water and filter into the same volumetric flask. Wash until acid-free. Make up to the mark.

Remarks:

3. Platinum crucibles should be cleaned as follows: Add to the crucibles a spoonful of potassium hydrogen sulphate. Heat on top of a blue flame until the salt has dissolved. Rotate the crucible while heating, making use of a crucible tong with platinum tips. Allow to cool and dissolve the salt in warm water. Rinse with water to remove the salt completely.
Note: Alsint crucibles should be cleaned with hot nitric acid.
4. With Alsint crucibles the HCl procedure should be finished rapidly since the liquid "creeps" over the rim of the crucible.
5. Hydrofluoric acid is a treacherous skin poison. Work in a good fume hood and wear rubber gloves, safety goggles and protective cloths. If any HF comes into contact with the skin, wash immediately and thoroughly with water and thereafter dab with 4 M ammonia (4.3) or calcium gluconate gel.
6. The calibration solutions for analysis have to be prepared in the same final medium as the samples in order to get a matrix, which is the same as in the samples. Since the addition of HF is evaporated the final medium is 0.12 M HCl.

# 6. REFERENCES

6.1 Membres du C.I.I.. 1969. Méthodes de référence pour la détermination des éléments minéraux dans les végétaux. L'Agronomie Tropicale 24: 827-835 (in French).

# 2.8 DIGESTION BY DRY-ASHING IN THE PRESENCE OF CAO

## 1. FIELD OF APPLICATION

1.1 This digestion is intended for the spectrometric determination of B.

## 2. PRINCIPLE

2.1 The organic plant material matrix is destroyed by controlled heating. Boron compounds are kept by calcium oxide.

## 3. APPARATUS

3.1 Porcelain crucibles.

3.2 Hot plate.

3.3 Muffle furnace.

3.4 Polythene stirring rods.

## 4. REAGENTS

4.1 Calcium Oxide Finely Powdered.

4.2 Sulphuric Acid Solution 0.5 mol/L - Add carefully, while swirling, 28 mL of concentrated sulphuric acid (96 %) to about 400 mL water in a 1-L volumetric flask. Let cool down and dilute to volume.

## 5. PROCEDURE

5.1 Weigh, to the nearest 0.001 g, approximately 1 g of the dried plant material sample in a porcelain crucible. Add 100 mg of calcium oxide (4.1) and mix thoroughly until no CaO particles can be distinguished any more from the plant material. Prepare also a blank digestion.
Place the crucible on a preheated hot plate (about 200 °C) and heat until the sample has charred completely. Take care that the sample does not start burning. Cool the charred sample and place it in a muffle furnace. Raise the temperature to 500 °C and maintain this temperature for 1.5 h. Take the crucible out of the furnace and allow to cool.
Add 10 mL of sulphuric acid (4.2). Let stand for 30 min and stir several times during this period. Filter over coarse paper into a polythene tube.

Remarks:

1. With the precipitate of $CaSO_4$ that has been formed after the addition of sulphuric acid other compounds will coprecipitate so that interferences are diminished.
2. The calibration solutions for analysis have to be prepared in the same digestion solution as the samples in order to get a matrix, which is the same as in the samples. Since part of the sulphuric acid is consumed by the calcium oxide, the final medium is approximately 0.35 M $H_2SO_4$.

# 6. REFERENCES

6.1 Gupta, U.C. 1967. The boron determination of some plant materials as determined with and without adding CaO before ashing. Plant Soil 26: 202-204.

# 3. EXTRACTIONS

## 3.1 EXTRACTION WITH WATER

### 1. FIELD OF APPLICATION

1.1 This extraction is meant for the determination of chloride ($Cl^-$), nitrite-nitrogen ($NO_2^-$), nitrate-nitrogen ($NO_3^-$) and sulphate-sulphur ($SO_4^{2-}$).

### 2. PRINCIPLE

2.1 The anions that are "free" in the plant material are extracted as such by water.

### 3. APPARATUS

3.1 Metal weighing funnel (Figure 1).

3.2 Shaker, linear or end-over-end.

3.3 Filter Paper, Fine, Cl free.

Remark:

1. Dried plant material may easily stick to glass when the relative humidity of the air is low. These homemade weighing funnels (stainless steel or aluminium) do not show this effect.

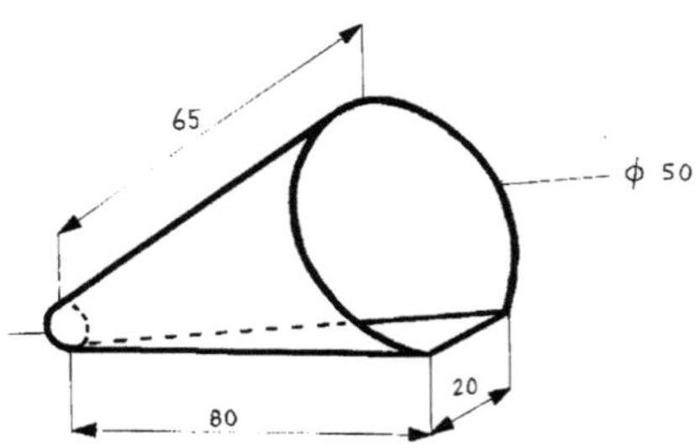


Figure 1 Weighing funnel (measures in mm).

### 4. PROCEDURE

4.1 Weigh, to the nearest 0.001 g, approximately 0.5 g of the dried plant material sample in a metal weighing funnel and transfer the sample to a 50-mL erlenmeyer flask. Add 25 mL water and shake for 30 min at room temperature. Filter over fine paper (4.1).

Collect the filtrate and filter this very portion of filtrate again over the same piece of paper. Use the last filtrate for analysis.

Remarks:

2. The first filtration will clog the paper just enough to obtain a more clear filtrate at the second time.
3. The determinations must be carried out on the same day as the extraction, since the filtrates will rapidly turn turbid due to mould growth.
4. The calibration solutions for analysis have to be prepared in the same final medium as the samples in order to get a matrix, which is the same as in the samples. The final medium is water.

# 3.2 EXTRACTION WITH HF - HCL

## 1. FIELD OF APPLICATION

1.1 This extraction is only appropriate for the determination of boron (B) and silicon (Si).

## 2. PRINCIPLE

2.1 Borates and silicates are extracted from the plant material as tetrafluoroborates and hexafluorosilicates, respectively.

Remarks:

1. If a sample contains both Si and K in large concentrations, the slightly soluble $K_2SiF_6$ may be formed. A larger extraction ratio should then be used.
2. The extracts are coloured brown, so that spectrometric determinations are not possible.
3. Only polythene flasks etc. can be used because glass contains Si.

## 3. APPARATUS

3.1 End-over-end shaker.

3.2 Polycarbonate test tubes and caps.

3.3 Polythene funnels.

3.4 Polythene volumetric flasks.

3.5 Polythene graduated cylinders.

3.6 Pipettors with polythene tips.

## 4. REAGENTS

4.1 Hydrofluoric Acid, 40 % (w/w), 22.6 mol/L ($\rho$ = 1.13 g/cm$^3$).

4.2 Hydrochloric Acid, 36 % (w/w), 12.0 mol/L ($\rho$ = 1.19 g/cm$^3$).

4.3 Extraction Mixture - Mix 100 mL of hydrofluoric acid (4.1) with 40 mL of hydrochloric acid (4.2) and add 360 mL water. This extraction solution is 4.5 M in HF and 1 M in HCl.

4.4 Ammonia Solution 4 mol/L - Dilute 75 mL of concentrated aqueous ammonia (25 %) with water to 250 mL. See remark 5.

# 5. PROCEDURE

5.1 Weigh, to the nearest 0.001 g, approximately 0.3 g of the dried plant material sample into a polycarbonate test tube. If Si has to be measured, 0.15 g $K_2SiF_6$ can be added as a check (see remark 1). Add 10 mL of extraction mixture (4.3) and moisten the plant material by shaking manually. Prepare also a blank extraction.
Shake overnight at room temperature. Filter the suspension over medium fine paper.

Remarks:

4. Since normal demineralised water may contain varying concentrations of Si, the use of ultra pure water is strongly advised.
5. Hydrofluoric acid is a treacherous skin poison. Work in a good fume hood and wear rubber gloves, safety goggles and protective cloths. If any HF comes into contact with the skin, wash immediately and thoroughly with water and thereafter dab with 4 M ammonia (4.4) or calcium gluconate gel.
6. Laboratory coats fresh from the laundry might release boron (from the perborate used in washing powder).
7. The calibration solutions for analysis have to be prepared in the same final medium as the samples in order to get a matrix, which is the same as in the samples. The final medium is 1.0 M HCl and 4.5 M HF.

# 6. REFERENCES

6.1 Novozamsky, I., R. van Eck and V.J.G. Houba. 1984. A rapid determination of silicon in plant material. Commun.Soil Sci.Plant Anal. 15: 205-211.

6.2 Van der Lee, J.J., I. Walinga, P.K. Manyeki, V.J.G. Houba and I. Novozamsky. 1987. Determination of boron in fresh and in dried plant material by plasma emission spectrometry after extraction with HF-HCl. Commun. Soil Sci. Plant Anal. 18: 789-802.

# 4. DETERMINATIONS

## 4.1 DETERMINATION OF ALUMINIUM

## 4.1.A DETERMINATION OF ALUMINIUM BY SPECTROPHOTOMETRY

### 1. PRINCIPLE OF METHOD

1.1 Aluminium ions form – in weakly acid medium and in the presence of a polycyclic ketoamine – an extremely stable, red-coloured complex with Eriochrome Cyanine R. The intensity of the red-coloured complex can be measured with a spectrophotometer at a wavelength of 595 nm.

1.2 This determination may be carried out on digest 2.3 (microwave digestion with $HNO_3$ - $H_2O_2$ - HF), digest 2.4 (digestion with $HNO_3$ - $H_2O_2$ - HF) and digest 2.7 (digestion by dry-ashing followed by treatment with HF).

### 2. RANGE AND DETECTION LIMIT

2.1 The procedure yields a standard curve that is linear up to approximately 5 mg/L Al.

2.2 The detection limit is approximately 20 μg/L in the digest. The determination limit is approximately 60 μg/L (3.0 mg/kg in the dried plant material).

### 3. INTERFERENCES

3.1 The polycyclic ketamine reagent is specific for Al; the only possible interference comes from Fe(III), but only at concentrations which are normally not found in plant material. Any Fe interference is prevented by addition of ascorbic acid.

### 4. PRECISION AND ACCURACY

4.1 The reproducibility of determinations by this procedure should give, at reasonable control and thorough sample preparation, a coefficient of variation within 8 %.

### 5. APPARATUS

5.1 Spectrophotometer.

### 6. REAGENTS

6.1A <u>Stock Solution, Al concentration 1000 mg/L</u> - Merck nr 1.19770.

6.1B Stock Solution, Al concentration 1000 mg/L - Dissolve 17.5760 g potassium aluminium sulphate dodecahydrate, $KAl(SO_4)_2.12H_2O$, in a volumetric flask of 1000 mL. Make up to 1000 mL with water.

6.2 Standard Solution, Al concentration 50 mg/L - Pipette 50.0 mL stock solution (6.1A or 6.1B) in a polythene 1000-mL volumetric flask and make up to volume with water.

6.3 Polycyclic Ketoamine - A heterocyclic fatty acid amine, to be purchased from Amchem Products, Inc., Ambler, Pa, USA, or from N.V. Mavom, Handelsweg 6, Alphen a/d Rijn, The Netherlands.

6.4 Ammonium Nitrate Solution 1.56 mol/L - Dissolve 6.25 g of ammonium nitrate, $NH_4NO_3$, in 50 mL water.

6.5 Sodium Chloride Solution 2.14 mol/L - Dissolve 6.25 g of sodium chloride, NaCl, in 50 mL water.

6.6 Eriochrome Cyanine R Solution - Dissolve 250 mg of Eriochrome Cyanine R (C.I. nr.43820) in 50 mL water. Add 50 mL of ammonium nitrate solution (6.4) and mix. Add 50 mL of sodium chloride solution (6.5) and mix. Add 2 mL of concentrated nitric acid (65 %) and mix. Make up to 250 mL with water.

6.7 Diluted Eriochrome Cyanine R Solution - Add 60 mL water to 40 mL of the Eriochrome Cyanine R solution (6.6).

6.8 Ascorbic Acid Solution - Dissolve 1 g of ascorbic acid, $C_6H_8O_6$, in 50 mL water. Prepare fresh daily.

6.9 Sulphite Solution 0.20 mol/L - Dissolve 1.25 g of sodium sulphite, $Na_2SO_3$, (or 2.5 g of sodium sulphite heptahydrate) in 50 mL water. Prepare fresh daily.

6.10 Polycyclic Ketoamine Solution - Dissolve 0.7 g of polycyclic ketoamine (6.3) in 100 mL water. Prepare fresh daily.

6.11 Acetate Solution 2.94 mol/L - Dissolve 20 g of sodium acetate trihydrate, $CH_3COONa .3H_2O$, in water and make up to 50 mL.

6.12 Mixed Reagent - Mix 50 mL of sulphite solution (6.9) with 50 mL of polycyclic ketoamine solution (6.10) and 50 mL of acetate solution (6.11). Add 225 mL water and mix. Prepare fresh daily. Do not use rubber stoppers.

Remark:

1. $KAl(SO_4).12H_2O$ may lose crystal water on standing. The reagent should be standardised by adding an excess of EDTA with $ZnSO_4$ and back-titration with Eriochrome Black T as an indicator.

## 7. CALIBRATION AND STANDARDS

7.1 Standard Series - Pipette 0 – 2.00 – 4.00 – 6.00 – 8.00 – 10.00 mL of the standard solution (6.2) into 100-mL volumetric flasks which already contain 40 mL water and add either 10 mL concentrated nitric acid (65 %) (digestion 2.3), 3.0 mL concentrated nitric acid (65 %) (digestion 2.4) or 1.0 mL concentrated hydrochloric acid (65 %) (digestion 2.7). Make up to the mark with water. Transfer the solutions to polythene bottles. This standard series has Al concentrations of 0 – 1 – 2 – 3 – 4 – 5 mg/L.

7.2 Calibration Curve - The absorbance (A) is plot versus mg/L aluminium in the standard series.

## 8. PROCEDURE

8.1 Measurement - Pipette 0.20 mL of the standard series, the blanks and the sample digests into test tubes. Add successively:

- 1.0 mL ascorbic acid solution (6.8); mix and wait for 5 min
- 1.0 mL diluted Eriochrome Cyanine R solution (6.7); mix and wait for 5 min
- 3.0 mL mixed reagent (6.12); mix and wait for 1 h

Measure the absorbance in a 1-cm cuvette at a wavelength of 595 nm.

Remarks:

1. The colour of the measuring solution is stable for 2 h.
2. All glassware must be pretreated with "chromic acid", to prevent Al desorption and to cleanse it from polycyclic ketoamine. Do not use a detergent, since this is contaminated by aluminium.
3. The pH of the final solution should lie between 5.5 and 5.1. The absorbance will increase with pH by 0.02 units per pH unit.

## 9. CALCULATION

9.1 The aluminium content of the dried plant material, expressed in mg/kg Al, is calculated by:

(a - b) * V / w

in which:

a is the concentration of aluminium in the sample digest, in mg/L;
b is the concentration of aluminium in the blank digest, in mg/L;
V is the total volume of digest at the end of the digestion procedure, in mL;
w is the weight of plant material sample, in g.

## 10. REFERENCES

10.1 Hill, U.T. 1966. New direct spectrophotometric determination of aluminium in steel, spelter, and iron ores. Anal. Chem. 38: 654 - 656.

# 4.1.B DETERMINATION OF ALUMINIUM BY ICP-OES

## 1. PRINCIPLE OF THE METHOD

1.1 Solutions with aluminium compounds are nebulised into an argon plasma, where all components are vaporised. Aluminium compounds are dissociated and excited, and then emit radiation of which the intensity is measured at a wavelength of 237.312 nm.

1.2 This determination may be carried out on digest 2.3 (microwave digestion with $HNO_3$ - $H_2O_2$ - HF), digest 2.4 (digestion with $HNO_3$ - $H_2O_2$ - HF) and digest 2.7 (digestion by dry-ashing followed by treatment with HF).

## 2. RANGE AND DETECTION LIMIT

2.1 This procedure yields a standard curve that is linear up to at least 100 mg/L Al.

2.2 The detection limit is approximately 0.035 mg/L in the digest. The determination limit is approximately 0.10 mg/L (12 respectively 33 mg/kg Al in the dried plant material).

## 3. INTERFERENCES

3.1 No interferences are expected.

## 4. PRECISION AND ACCURACY

4.1 The reproducibility of determinations by this procedure should give, at reasonable control and thorough sample preparation, a coefficient of variation within 5 % + 5 mg/kg.

## 5. APPARATUS

5.1 Inductively coupled plasma optical emission spectrometer.

## 6. REAGENTS

6.1A Stock Solution, Al concentration 1000 mg/L - Merck nr 1.19770.

6.1B Stock Solution, Al concentration 1000 mg/L - Dissolve 17.5760 g potassium aluminium sulphate dodecahydrate, $KAl(SO_4)_2.12H_2O$, in a volumetric flask of 1000 mL. Make up to 1000 mL with water.

Remark:

1. $KAl(SO_4)_2.12H_2O$ may lose crystal water on standing. The reagent should be standardised by adding an excess of EDTA and back-titration of the excess EDTA with $ZnSO_4$ solution with Eriochrome Black T as an indicator.

## 7. CALIBRATION AND STANDARDS

7.1 Standard Series - Pipette 0 – 1.00 – 2.00 – 10.00 mL of the standard solution (6.1) into 100-mL volumetric flasks which already contain 40 mL water and add either 10 mL concentrated nitric acid (65 %) (digestion 2.3), 3.0 mL concentrated nitric acid (65 %) (digestion 2.4) or 1.0 mL concentrated hydrochloric acid (65 %) (digestion 2.7). Let cool down and make up to the mark with water. This standard series has Al concentrations of 0 – 10 – 20 – 100 mg/L.

Remark:

2. A mix standard series can be used for simultaneous measurements with ICP-OES (Al, Ca, Cd, Co, Cr Cu, Fe, K, Mg, Mn, Na, Ni P, S and Zn).

7.2 Calibration Curve - The emission counts are plot versus mg/L aluminium in the standard series.

## 8. PROCEDURE

8.1 Measurement - Measure in the standard series, the blanks and the sample digests the Al concentration with the ICP-OES at a wavelength of 237.312 nm. At this wavelength a (fitted) background correction is used.

Remarks:

3. A data management system and system controller are used. In this way the measurements are checked continuously and the data output is in concentration units in the digests.
4. Due to the linearity of the ICP-OES it is possible to calibrate the ICP-OES only with the highest and zero standard of the standard series. As a check all standards can be measured as samples.
5. Scandium (5 mg/L), at a wavelength of 255.235 nm, is used in the author's laboratory as an internal standard to compensate for matrix effects.

## 9. CALCULATION

9.1 The total aluminium content in the dried plant material, expressed in mg/kg Al, is calculated by:

$(a - b) * V / w$

in which:

a is the concentration of aluminium in the sample digest, in mg/L;
b is the concentration of aluminium in the blank digest, in mg/L;
V is the total volume of digest at the end of the digestion procedure, in mL;
w is the weight of plant material sample, in g.

# 4.1.C DETERMINATION OF ALUMINIUM BY ICP-MS

## 1. PRINCIPLE OF THE METHOD

1.1 Solutions with aluminium compounds are nebulised into an argon plasma, where all components are vaporised. The ions produced are entrained in the plasma gas and introduced into a mass spectrometer, sorted according to their mass-to-charge ratios, and quantified with a channel electron multiplier. Aluminium is determined at mass 27 amu.

1.2 This determination may be carried out on digest 2.3 (microwave digestion $HNO_3$ - $H_2O_2$ - HF), digest 2.4 (digestion with $HNO_3$ - $H_2O_2$ - HF) and digest 2.7 (digestion by dry-ashing followed by treatment with HF).

## 2. RANGE AND DETECTION LIMIT

2.1 This procedure yields a standard curve that is linear up to at least 1000 μg/L Al.

2.2 The detection limit is approximately 0.4 μg/L in the digest. The determination limit is approximately 1.2 μg/L (1.3 respectively 4 mg/kg in the dried plant material).

## 3. INTERFERENCES

3.1 Interferences are expected from $^{54}Fe^{++}$ and $^{11}B^{16}O$ due to mass overlap with $^{27}Al$.

## 4. PRECISION AND ACCURACY

4.1 The reproducibility of determinations by this procedure should give, at reasonable control and thorough sample preparation, a coefficient of variation within 10 %.

## 5. APPARATUS

5.1 Inductively coupled plasma mass spectrometer.

## 6. REAGENTS

6.1A <u>Stock Solution, Al concentration 1000 mg/L</u> - Merck nr 1.19770.

6.1B <u>Stock Solution, Al concentration 1000 mg/L</u> - Dissolve 17.5760 g potassium aluminium sulphate dodecahydrate, $KAl(SO_4)_2.12H_2O$, in a volumetric flask of 1000 mL. Make up to 1000 mL with ultra pure water.

6.2 <u>Standard Solution, Al concentration 20 mg/L</u> - Pipette 1 mL concentrated nitric acid (65 % s.p.) in a 100-mL polythene volumetric flask which already contains about 50 mL ultra pure water. Add 2.00 mL stock solution (6.1A or 6.1B) and make up to volume with ultra pure water.

Remark:

1. $KAl(SO_4).12H_2O$ may lose crystal water on standing. The reagent should be standardised by adding an excess of EDTA with $ZnSO_4$ and back-titration with Eriochrome Black T as an indicator.

## 7. CALIBRATION AND STANDARDS

7.1 Standard Series - pipette 0 – 1.00 – 2.00 – 5.00 mL of the standard solution (6.2) in a 100-mL polythene volumetric flasks which already contain 40 mL ultra pure water. Add either 1.0 mL concentrated nitric acid (65 %) (digestion 2.3), 0.3 mL concentrated nitric acid (65 %) (digestion 2.4) or 0.1 mL concentrated hydrochloric acid (36 %) (digestion 2.7). Let cool down and make up to the mark with ultra pure water. This standard series has Al concentrations of 0 – 200 – 400 – 1000 µg/L.

7.2 Calibration Curve - The counts per second are plot versus µg/L aluminium in the standard series.

Remark:

2. A mix standard series can be used for simultaneous measurements with ICP-MS (Al, As, Cd, Co, Cr, Cu, Mn, Ni, Pb, Sb, Sn, V and Zn).

## 8. PROCEDURE

8.1 Measurement – Dilute the blanks and the sample digests 1 + 9 (v/v) with ultra pure water and mix. Measure in the standard series, the blanks and the sample digests the Al concentration with the ICP-MS at a mass of 27 amu. Make use of corrections if necessary.

Remarks:

3. A data management system and system controller are used. In this way the measurements are checked continuously and the data output is in concentration units in the digests.
4. Due to the linearity of the ICP-MS it is possible to calibrate the ICP-MS only with the highest and zero standard of the standard series. As a check all standards can be measured as samples.

## 9. CALCULATION

9.1 The total aluminium content in the dried plant material, expressed in mg/kg Al, is calculated by:

$0.01 * (a - b) * V / w$

in which:

a is the concentration of aluminium in the diluted sample digest, in µg/L;

b is the concentration of aluminium in the diluted blank digest, in µg/L;

V is the total volume of digest at the end of the digestion procedure, in mL;

w is the weight of plant material sample, in g.

# 4.2 DETERMINATION OF ARSENIC

## 4.2.A DETERMINATION OF ARSENIC BY ICP-MS

### 1. PRINCIPLE OF THE METHOD

1.1 Solutions with arsenic compounds are nebulised into an argon plasma, where all components are vaporised. The ions produced are entrained in the plasma gas and introduced into a mass spectrometer, sorted according to their mass-to-charge ratios, and quantified with a channel electron multiplier. Arsenic is determined at mass 75 amu.

1.2 This determination may be carried out on digest 2.3 (microwave digestion with $HNO_3$ - $H_2O_2$ - HF), digest 2.4 (digestion with $HNO_3$ - $H_2O_2$ - HF) and digest 2.7 (digestion by dry-ashing followed by treatment with HF).

### 2. RANGE AND DETECTION LIMIT

2.1 This procedure yields a standard curve that is linear up to at least 50 µg/L As.

2.2 The detection limit is approximately 0.007 µg/L in the digest. The determination limit is approximately 0.021 µg/L (2 respectively 7 µg/kg in the dried plant material).

### 3. INTERFERENCES

3.1 Interferences are expected from $^{40}Ar^{35}Cl$, $^{40}Ca^{35}Cl$, $^{63}Cu^{12}C$ due to mass overlap at $^{75}As$. A correction factor for ArCl can be used. The correction affects also the detection limit.

### 4. PRECISION AND ACCURACY

4.1 The reproducibility of determinations by this procedure should give, at reasonable control and thorough sample preparation, a coefficient of variation within 10 %.

### 5. APPARATUS

5.1 Inductively coupled plasma mass spectrometer.

### 6. REAGENTS

6.1 Stock Solution, As concentration 1000 mg/L - Merck nr 1.02602.

6.2 Standard Solution, As concentration 1 mg/L - Pipette 1 mL concentrated nitric acid (65 % s.p.) in a 1000-mL polythene volumetric flask which already contains about

500 mL ultra pure water. Add 1.00 mL stock solution (6.1) and make up to volume with ultra pure water.

## 7. CALIBRATION AND STANDARDS

7.1 Standard Series - pipette 0 – 1.00 – 2.00 – 5.00 mL of the standard solution (6.2) in a 100-mL polythene volumetric flasks which already contain 40 mL ultra pure water. Add either 1.0 mL concentrated nitric acid (65 %) (digestion 2.3), 0.3 mL concentrated nitric acid (65 %) (digestion 2.4) or 0.1 mL concentrated hydrochloric acid (36 %) (digestion 2.7). Let cool down and make up to the mark with ultra pure water. This standard series has As concentrations of 0 – 10 – 20 – 50 µg/L.

7.2 Calibration Curve - The counts per second are plot versus µg/L arsenic in the standard series.

Remark:

1. A mix standard series can be used for simultaneous measurements with ICP-MS (Al, As, Cd, Co, Cr, Cu, Mn, Ni, Pb, Sb, Sn, V and Zn).

## 8. PROCEDURE

8.1 Measurement - Measure in the standard series, the blanks and sample digests the As concentration with the ICP-MS at a mass of 75 amu. Make use of corrections if necessary.

Remarks:

2. A data management system and system controller are used. In this way the measurements are checked continuously and the data output is in concentration units in the digests.
3. Due to the linearity of the ICP-MS it is possible to calibrate the ICP-MS only with the highest and zero standard of the standard series. As a check all standards can be measured as samples.

## 9. CALCULATION

9.1 The total arsenic content in the dried plant material, expressed in µg/kg As, is calculated by:

$(a - b) * V / w$

in which:

a is the concentration of arsenic in the sample digest, in µg/L;
b is the concentration of arsenic in the blank digest, in µg/L;
V is the total volume of digest at the end of the digestion procedure, in mL;
w is the weight of plant material sample, in g.

# 4.3 DETERMINATION OF BORON

# 4.3.A DETERMINATION OF BORON BY SPECTROPHOTOMETRY

## 1. PRINCIPLE OF METHOD

1.1 Borate ions form a red coloured compound with Azomethine-H at pH 4-5. The absorbance of the complex is measured at a wavelength of 430 nm.

1.2 This determination may be carried out on digest 2.8 (digestion by dry-ashing in the presence of CaO).

## 2. RANGE AND DETECTION LIMIT

2.1 This procedure yields a standard curve that is linear up to approximately 2 mg/L B.

2.2 The detection limit is approximately 0.1 mg/L in the digest. The determination limit is approximately 0.3 mg/L (3 mg/kg dried plant material).

## 3. INTEFERENCES

3.1 Al, Cu and Fe may interfere.

<u>Remarks:</u>

1. Interference from Al and Cu is prevented by addition of EDTA.
2. Thioglycollic acid is used to mask Fe.

## 4. PRECISION AND ACCURACY

4.1 The reproducibility of determinations by this procedure should give, at reasonable control and thorough sample preparation, a coefficient of variation within 10 %.

## 5. APPARATUS

5.1 Polythene test tubes with covers (because of bad smell).

5.2 Spectrophotometer with interference filter.

## 6. REAGENTS

6.1A <u>Stock Solution, B Concentration 1000 mg/L</u> - Merck nr 1.19500.

6.1B Stock Solution, B Concentration 1000 mg/L - Dissolve 8.819 g of borax (sodium tetraborate decahydrate), $Na_2B_4O_7.10H_2O$, (see remark 3) in some water in a 1000-mL volumetric flask and make up to volume. Store in a polythene bottle.

6.2 Standard Solution, B concentration 50 mg/L - Dilute 5.00 mL of the stock solution (6.1A or 6.1B) with water to 100 mL. Store in a polythene bottle.

6.3 Thioglycollic Acid, 98 % (Mercaptoacetic Acid, $C_2H_4O_2S$, $\rho = 1.32$ g/cm$^3$).

6.4 Azomethine-H (4-Hydroxy-5[Salicylidene-Amino]-2,7-Naphtalene Disulphonic Acid, $C_{17}H_{13}NO_8S_2$).

6.5 Azomethine-H Mixture - Dissolve 2 g of ascorbic acid and 0.9 g of Azomethine-H (6.4) in some water and make up to 100 mL in a polythene flask. Prepare fresh daily.

6.6 Mixed Buffer Solution, pH 4.9 - Dissolve 100 g of ammonium acetate, $CH_3COONH_4$, in 160 mL water. Add 50 mL of concentrated acetic acid (100 %) and mix. Dissolve 2.68 g of $Na_2EDTA.2H_2O$ in the mixture. Add 2.4 mL of thioglycollic acid (6.3), mix well and allow to stand overnight. Store in a polythene bottle. Prepare fresh weekly.

Remarks:

3. Borax easily loses crystal water on standing. To be sure that the salt contains 10 molecules of crystal water it has to be recrystallised. First dissolve 30 g borax 100 mL water, then filter the newly formed crystals on a Büchner funnel by suction. Wash twice with 15 mL water, 15 mL ethanol and finally with 15 mL of diethyl ether. Each washing must be followed by suction to remove the liquid. After the final washing, the solid has to be dried on a watch glass at room temperature for 12-18 h (Besset, et al, 1978).
4. To avoid any problems with borax, the use of boric acid, $H_3BO_3$, may be considered. This compound should be used as such, because drying at 105 °C, or even storage in a desiccator, causes loss of water and conversion to metaboric acid, $HBO_2$. A stock solution of 100 mg/L B will contain 0.5719 g $H_3BO_3$ per litre.
5. To prevent desorption of borate from the glassware to be used, this must be filled with 4 M $HNO_3$ and left overnight before cleaning in the usual manner. For the same reason, reagents and standard solutions should be transferred to polythene vessels directly after preparation.
6. The mixed buffer solution (6.6) should be used only after standing overnight, since the colour to be measured fades when using the freshly prepared buffer. It is assumed that the thioglycollic acid (6.3) is responsible for this effect.

## 7. CALIBRATION AND STANDARDS

7.1 Standard series - Pipette 0 – 0.50 – 1.00 – 2.00 – 3.00 – 4.00 mL of the standard solution (6.2) into 100-mL volumetric flasks which already contain 40 mL water and add 2.0 mL concentrated sulphuric acid (96 %) (digestion 2.8). Let cool down and make up to volume with water. Store in polythene bottles. This standard series has B concentrations of 0 – 0.25 – 0.50 – 1.0 – 1.5 – 2.0 mg/L.

7.2 Calibration Curve - The absorbance (A) is plot versus mg/L boron in the standard series.

# 8. PROCEDURE

8.1 Pipette 2.00 mL of the standard series, the blanks and the sample digests into polythene test tubes. Add 4.0 mL of mixed buffer solution (6.6) and mix thoroughly. Add 2.0 mL of Azomethine-H mixture (6.5) and mix again thoroughly. Measure the absorbance in a 1-cm cuvette at a wavelength of 430 nm after 30 min and within 1.5 h.

Remarks:

7. The calibration curve should be linear. The zero standard normally reads 0.190 - 0.225 Abs and the highest standard usually reads 0.550 - 0.590 Abs.
8. Instead of the 410 nm mentioned in literature, 430 nm was chosen for absorbance measurement because the background absorption of the reagent is much lower at this wavelength.
9. Laboratory coats fresh from the laundry might release boron (from the perborate used in washing powder).

# 9. CALCULATION

9.1 The boron content of the dried plant material, expressed in mg/kg B, is calculated by:

$(a - b) * V / w$

in which:

a is the concentration of boron in the sample digest, in mg/L;

b is the concentration of boron in the blank digest, in mg/L;

V is the total volume of digest at the end of the digestion procedure, in mL;

w is the weight of plant material sample, in g.

# 10. REFERENCES


10.1 Wolf, B. 1971. The determination of boron in plant material extracts, plant materials, composts, manures, water and nutrient solutions. Commun. Soil Sci. Plant Anal. 2: 363-374.

10.2 De Bes, S.S. 1973. P.A. den Dekker and P.A. van Dijk. The determination of boron with Azomethine-H in plant material, soil and water. Internal communication of the Glasshouse Crops Research Station, Naaldwijk, The Netherlands (in Dutch).

10.3 Wolf, B. 1974. Improvement of the Azomethine-H method for the determination of boron. Commun. Soil Sci. Plant Anal. 5: 39-44.

10.4 Besset, J et al. 1978. Vogel's textbook of quantitative inorganic analysis, 4th edition, Longman, London. p. 300.

# 4.3.B DETERMINATION OF BORON BY ICP-OES

## 1. PRINCIPLE OF THE METHOD

1.1 Solutions with boron compounds are nebulised into an argon plasma, where all components are vaporised. Boron compounds are dissociated and excitated, and then emit radiation of which the intensity is measured at a wavelength of 249.678 nm.

1.2 This determination may be carried out on extract 3.2 (extraction with HF - HCl).

## 2. RANGE AND DETECTION LIMIT

2.1 This procedure yields a standard curve that is linear up to at least 10 mg/L B.

2.2 The detection limit is approximately 0.002 mg/L in the extract. The determination limit is approximately 0.006 mg/L (0.02 respectively 0.06 mg/kg in the dried plant material).

## 3. INTERFERENCES

3.1 High concentrations of Fe cause spectral interference.

## 4. PRECISION AND ACCURACY

4.1 The reproducibility of determinations by this procedure should give, at reasonable control and thorough sample preparation, a coefficient of variation within 10 % + 10 mg/kg.

## 5. APPARATUS

5.1 End-over-end shaker.

5.2 Polycarbonate test tubes.

5.3 Polythene bottles.

5.4 Polythene volumetric flasks.

5.5 Inductively coupled plasma optical emission spectrometer equipped with a polythene nebuliser.

# 6. REAGENTS

6.1A Stock Solution. B concentration 1000 mg/L - Merck nr 1.19500.

6.1B Stock Solution. B concentration 1000 mg/L Dissolve 8.819 g of borax (sodium tetraborate decahydrate), $Na_2B_4O_7.10H_2O$, (see remark 1) in some water in a 1000-mL volumetric flask and make up to volume. Store in a polythene bottle.

Remark:

1. Borax easily loses crystal water on standing. To be sure that the salt contains 10 molecules of crystal water it has to be recrystallised. First dissolve 30 g borax in 100 mL water, then filter the newly formed crystals on a Büchner funnel by suction. Wash twice with 15 mL water, 15 mL ethanol and finally with 15 mL of diethyl ether. Each washing must be followed by suction to remove the liquid. After the final washing, the solid has to be dried on a watch glass at room temperature for 12-18 h (Besset et al, 1978).

# 7. CALIBRATION AND STANDARDS

7.1 Standard Series - Pipette 0 – 0.50 – 1.00 mL of the stock solution (6.1) into 100-mL polythene volumetric flasks which already contain 40 mL water and add 20 mL concentrated hydrofluoric acid (40 %) and 8.5 mL concentrated hydrochloric acid (36 %) (extraction 3.2). Let cool down and make up the mark with water. This standard series has B concentrations of 0 – 5.0 – 10.0 mg/L.

7.2 Calibration Curve - The emission counts are plot versus mg/L boron in the standard series.

Remark:

2. The pipetting should be done by means of piston-type pipettes with polythene tips.
3. A mix standard series can be used for simultaneous measurements with ICP-OES (B and Si).

# 8. PROCEDURE

8.1 Measurement - Measure in the standard series, the blanks and the sample extracts the B concentration with the ICP-OES at a wavelength of 249.772 nm. At this wavelength a (fitted) background correction is used.

Remarks:

4. Beryllium (5 mg/L), at a wavelength of 234.861 nm, is used in the author's laboratory as an internal standard to compensate for matrix effects.
5. Laboratory coats fresh from the laundry might release boron (from the perborate used in washing powder).

# 9. CALCULATION

9.1 The total boron content in the dried plant material, expressed in mg/kg B, is calculated by:

(a - b) * V / w

in which:
a is the concentration of boron in the sample digest, in mg/L;
b is the concentration of boron in the blank digest, in mg/L;
V is the total volume of digest at the end of the digestion procedure, in mL;
w is the weight of plant material sample, in g.

## 10. REFERENCES

10.1 Novozamsky, I., R. van Eck, V.J.G. Houba, and J.J. van der Lee. 1990. A new solvent extraction for the determination of traces of boron by ICP-AES. Atomic Spectrosc. 11: 83-84.

10.2 Besset, J et al. 1978. Vogel's textbook of quantitative inorganic analysis, 4th edition, Longman, London. p. 300.

# 4.4 DETERMINATION OF CALCIUM

# 4.4.A DETERMINATION OF CALCIUM BY FLAME AES

## 1. PRINCIPLE OF THE METHOD

1.1 The sample is nebulised into an air-acetylene flame, where it is vaporised; calcium compounds are atomised and the CaOH molecules thus formed emit radiation of which the intensity is measured at a wavelength of 622.0 nm.

Remark:

1. Although Ca measurements are also possible at 423 nm and 554 nm, the spectral interferences at 622 nm are smaller for elements that are present in plant material, whereas larger interferences at this wavelength only come from elements that hardly or not occur in plant material.

1.2 This determination may be carried out on digest 2.1 (digestion with $H_2SO_4$ - salicylic acid - $H_2O_2$ - Se), digest 2.2 (digestion with $H_2SO_4$ - salicylic acid - $H_2O_2$), digest 2.3 (microwave digestion with $HNO_3$ - $H_2O_2$ - HF), digest 2.4 (digestion with $HNO_3$ - $H_2O_2$ - HF) and digest 2.7 (digestion by dry-ashing followed by treatment with HF).

## 2. RANGE AND DETECTION LIMIT

2.1 This procedure yields a standard curve that is linear up to approximately 50 mg/L Ca.

2.2 The detection limit is approximately 2 mg/L in the digest. The determination limit is approximately 6 mg/L (7-25 mmol/kg in the dried plant material).

2.3 If the calcium content for digestion 2.1 and 2.2 is higher than 1000 mmol/kg dry material, the solubility product of $CaSO_4$ can be surpassed in the digest; a smaller sample has to be weighed out then.

## 3. INTERFERENCES

3.1 When using the cooler air-propane flame the emission signal is 30 - 40 % lower, and there will be more interferences.

3.2. When an air-acetylene flame is used, there may be interferences from P and Al, by forming stable $Ca_2(PO_4)_3$ and $Ca_2Al_2O_4$ complexes in the flame. La can be used as a releasing agent to release Ca from the interfering compound.

## 4. PRECISION AND ACCURACY

4.1 The reproducibility of determinations by this procedure should give, at reasonable control and thorough sample preparation, a coefficient of variation within 5 % + 5 mmol/kg.

## 5. APPARATUS

5.1 Flame atomic emission spectrometer.

## 6. REAGENTS

61A Stock Solution, Ca concentration 1000 mg/L - Merck nr 1.19778.

6.1B Stock Solution, Ca concentration 1000 mg/L - Weigh out 2.497 g of calcium carbonate, $CaCO_3$ (dried at 105 °C for 2 hours, allow cooling in a desiccator). Transfer it to a 1000-mL volumetric flask with the help of about 150 mL water. Add 13 mL of 4 M hydrochloric acid (6.2) and boil to expel $CO_2$. If calcium carbonate particles remain visible, add 1 mL of 4 M hydrochloric acid (6.2) extra. Allow to cool and make up to volume with water.

6.2 Hydrochloric Acid 4 mol/L - Add 34 mL of concentrated hydrochloric acid (36 %) to about 400 mL water and make up to 1 litre.

6.3A Lanthanum Solution, La concentration 1.1 g/L - Dissolve 3.43 g lanthanum nitrate hexahydrate, $La(NO_3)_3.6H_2O$, in a 1000-ml volumetric flask and make up to the mark with water.

6.3B Cesium-Lanthanum Solution, Cs concentration 1.1 g/L, La concentration 1.1 g/L - Dissolve 1.4 g cesium chloride, CsCl, and 3.43 g lanthanum nitrate hexahydrate, $La(NO_3)_3.6H_2O$, in a 1000-ml volumetric flask and make up to the mark with water.

Remark:

2. The calcium carbonate has to be dried at 105 °C for at least 2 hours just before weighing. Just before drying, any lumps should be cut down so that only fine crystals remain. This is important, because coarse crystals contain much occluded water in their cavities and thus would need a higher drying temperature. The fine crystals should, however, not be pulverised because this enhances deliquescence (= attraction of water) and/or efflorescence (= weathering) later on. After the recommended heating procedure, the calcium carbonate should be cooled in a desiccator containing preferably magnesium perchlorate as a desiccant. The dried calcium carbonate should be weighed as soon as it has reached the ambient temperature.

3. The cesium chloride should be of the highest analytical grade ("pro analysi"), because a lower quality (e.g. "reinst" = "most pure") contains much more Na.

# 7. CALIBRATION AND STANDARD SERIES

7.1 Standard Series - Pipette 0 – 10.0 – 20.0 – 30.0 – 40.0 – 50.0 mL of the stock solution (6.1A or 6.1B) into 100-mL volumetric flasks which already contain 40 mL water. Add either 4.5 mL concentrated sulphuric acid (96 %) (digestion 2.1 or 2.2), 10 mL concentrated nitric acid (65 %) (digestion 2.3), 3.0 mL concentrated nitric acid (65 %) (digestion 2.4) or 1.0 mL concentrated hydrochloric acid (36 %) (digestion 2.7). Let cool down and make up to the mark with water. This standard series has Ca concentrations of 0 – 100 – 200 – 300 – 400 – 500 mg/L.

7.2 Calibration Curve - The emission counts are plot versus mg/L calcium in the standard series.

Remark:

4. A mix standard series can be used for simultaneous measurements with Flame-AES (Ca, K and Na). In this case Cs is necessary as an ionisation buffer for the determination of K (see section 4.11.A) or Na (see section 4.15.A).

# 8. PROCEDURE

8.1 Dilute the standard series, the blanks and sample digests 1 + 9 (v/v) with lanthanum - solution (6.3A) or with cesium - lanthanum solution (6.3B) (see remark 3) and mix. Measure in the diluted standard series, the diluted blanks and the diluted sample digests the Ca concentration with flame AES at a wavelength of 622.0 nm, using an air-acetylene flame.

Remark:

5. Instead of indicating a wavelength, a simple emission spectrometer ("flame photometer") may only be supplied with a so-called Ca filter. Make sure that this is an interference filter, since glass filters are not selective enough.

# 9. CALCULATION

9.1 The calcium content of the dried plant material, expressed in mmol/kg, is calculated by:

$0.02495 * (a - b) * V / w$

in which:

a is the concentration of calcium in the sample digest, in mg/L;

b is the concentration of calcium in the blank digest, in mg/L;

V is the total volume of digest at the end of the digestion procedure, in mL;

w is the weight of plant material sample, in g.

# 4.4.B DETERMINATION OF CALCIUM BY FLAME AAS

## 1. PRINCIPLE OF THE METHOD

1.1 The sample is nebulised into an air-acetylene flame, where it is vaporised; calcium compounds are atomised and the calcium atoms thus formed absorb radiation from a hollow-cathode lamp. The absorbance is measured at a wavelength of 422.7 nm.

1.2 This determination may be carried out on digest 2.1 (digestion with $H_2SO_4$ - salicylic acid - $H_2O_2$ - Se), digest 2.2 (digestion with $H_2SO_4$ - salicylic acid - $H_2O_2$), digest 2.3 (microwave digestion with $HNO_3$ - $H_2O_2$ - HF), digest 2.4 (digestion with $HNO_3$ - $H_2O_2$ - HF) and digest 2.7 (digestion by dry-ashing followed by treatment with HF).

## 2. RANGE AND DETECTION LIMIT

2.1 This procedure yields a standard curve that is linear up to approximately 5 mg/L Ca.

2.2 The detection limit is approximately 2 mg/L in the digest. The determination limit is approximately 6 mg/L (7-25 mmol/kg in the dried plant material).

2.3 If calcium contents for digestion 2.1 and 2.2 are higher than 1000 mmol/kg dry material, the solubility product of $CaSO_4$ can be surpassed in the digest; a smaller sample has to be weighed out then.

## 3. INTERFERENCES

3.1 Since Ca atoms are easily captured in the flame into poorly dissociating compounds (like oxides), a releasing agent like lanthanum must be added.

## 4. PRECISION AND ACCURACY

4.1 The reproducibility of determinations by this procedure should give, at reasonable control and thorough sample preparation, a coefficient of variation within 5 % + 5 mmol/kg.

## 5. APPARATUS

5.1 Flame atomic absorption spectrophotometer.

## 6. REAGENTS

6.1A Stock solution, Ca concentration 1000 mg/L - Merck nr 1.19778.

6.1B Stock solution, Ca concentration 1000 mg/L - Weigh out 2.497 g of $CaCO_3$ (dried at 105 °C for 2 hours, allow cooling in a desiccator) and transfer it with 100-150 mL water to a 1-litre volumetric flask. Add 13 mL of 4 M hydrochloric acid (6.2) and boil to expel $CO_2$. If calcium carbonate particles remain visible, add 1 mL of 4 M hydrochloric acid (6.2) extra. Allow to cool and make up to volume with water.

6.2 Hydrochloric Acid 4 mol/L - Add 34 mL of concentrated hydrochloric acid (36 %) to about 400 mL water and make up to 1 litre.

6.3 Lanthanum solution, La concentration 1 g/L - Dissolve 3.12 g of lanthanum nitrate hexahydrate, $La(NO_3)_3.6H_2O$, in some water and make up to 1 litre with water.

Remark:

1. The calcium carbonate has to be dried at 105 °C for at least 2 hours just before weighing. Just before drying, any lumps should be cut down so that only fine crystals remain. This is important, because coarse crystals contain much occluded water in their cavities and thus would need a higher drying temperature. The fine crystals should, however, not be pulverised because this enhances deliquescence (= attraction of water) and/or efflorescence (= weathering) later on. After the recommended heating procedure, the calcium carbonate should be cooled in a desiccator containing preferably magnesium perchlorate as a desiccant. The dried calcium carbonate should be weighed as soon as it has reached the ambient temperature.

## 7. CALIBRATION AND STANDARDS

7.1 Standard series - Pipette 0 – 2.00 – 4.00 – 6.00 – 8.00 – 10.00 mL of the stock solution (6.1A or 6.1B) into 100-mL volumetric flasks which already contain 40 mL water. Add either 4.5 mL concentrated sulphuric acid (96 %) (digestion 2.1 or 2.2), 10 mL concentrated nitric acid (65 %) (digestion 2.3), 3.0 mL concentrated nitric acid (65 %) (digestion 2.4) or 1.0 mL concentrated hydrochloric acid (36 %) (digestion 2.7). Let cool down and make up to the mark with water. This standard series has Ca concentrations of 0 – 20 – 40 – 60 – 80 – 100 mg/L.

7.2 Calibration Curve - The absorbance (A) is plot versus mg/L calcium in the standard series.

## 8. PROCEDURE

8.1 Dilute the standard series, the blanks and the sample digests 1 + 19 (v/v) with lanthanum nitrate solution (6.3) and mix. Measure in the diluted standard series, the diluted blanks and the diluted sample digests the Ca concentration with flame AAS at a wavelength of 422.7 nm, using a yellow (reducing) air-acetylene flame.

Remark:

2. The calibration curve should be linear by virtue of the releasing action of lanthanum.

# 9. CALCULATION

9.1 The calcium content of the dried plant material, expressed in mmol/kg Ca, is calculated by:

$0.02495 * (a - b) * V / w$

in which:

a is the concentration of calcium in the sample digest, in mg/L;
b is the concentration of calcium in the blank digest, in mg/L;
V is the total volume of digest at the end of the digestion procedure, in mL;
w is the weight of plant material sample, in g.

# 4.4.C DETERMINATION OF CALCIUM BY ICP-OES

## 1. PRINCIPLE OF THE METHOD

1.1 Solutions with calcium compounds are nebulised into an argon plasma, where all components are vaporised. Calcium compounds are dissociated and excited, and then emit radiation of which the intensity is measured at a wavelength of 317.933 nm.

1.2 This determination may be carried out on digest 2.1 (digestion with $H_2SO_4$ - salicylic acid - $H_2O_2$ - Se), digest 2.2 (digestion with $H_2SO_4$ - salicylic acid - $H_2O_2$), digest 2.3 (microwave digestion with $HNO_3$ - $H_2O_2$ - HF), digest 2.4 (digestion with $HNO_3$ - $H_2O_2$ - HF) and digest 2.7 (digestion by dry-ashing followed by treatment with HF).

## 2. RANGE AND DETECTION LIMIT

2.1 This procedure yields a standard curve that is linear up to at least 200 mg/L Ca.

2.2 The detection limit is approximately 0.004 mg/L in the digest. The determination limit is approximately 0.012 mg/L (0.03 respectively 0.1 mmol/kg in plant material).

## 3. INTERFERENCES

3.1 No interferences are expected.

## 4. PRECISION AND ACCURACY

4.1 The reproducibility of determinations by this procedure should give, at reasonable control and thorough sample preparation, a coefficient of variation within 5 % + 5 mmol/kg.

## 5. APPARATUS

5.1 Inductively coupled plasma optical emission spectrometer.

## 6. REAGENTS

6.1A Stock Solution, Ca concentration 1000 mg/L - Merck nr 1.19778.

6.1B Stock solution, Ca concentration 1000 mg/L - Weigh out 2.497 g of $CaCO_3$ (see remark 1) and transfer it with 100-150 mL water to a 1-litre volumetric flask. Add 13 mL of 4 M hydrochloric acid (6.2) and boil to expel $CO_2$. If calcium carbonate particles remain visible, add 1 mL of 4 M hydrochloric acid (6.2) extra. Allow to cool and make up to volume with water.

6.2 Hydrochloric Acid 4 mol/L - Add 34 mL of concentrated hydrochloric acid (36 %) to about 400 mL water and make up to 1 litre.

Remark:

1. The calcium carbonate has to be dried at 105 °C for at least 2 hours just before weighing. Just before drying, any lumps should be cut down so that only fine crystals remain. This is important, because coarse crystals contain much occluded water in their cavities and thus would need a higher drying temperature. The fine crystals should, however, not be pulverised because this enhances deliquescence (= attraction of water) and/or efflorescence (= weathering) later on. After the recommended heating procedure, the calcium carbonate should be cooled in a desiccator containing preferably magnesium perchlorate as a desiccant. The dried calcium carbonate should be weighed as soon as it has reached the ambient temperature.

## 7. CALIBRATION AND STANDARDS

7.1 Standard series - Pipette 0 – 1.00 – 2.00 – 20.00 mL of the stock solution (6.1A or 6.1B) into 100-mL volumetric flasks, which already contain 40 mL water. Add either 4.5 mL concentrated sulphuric acid (96 %) (digestion 2.1 or 2.2), 10 mL concentrated nitric acid (65 %) (digestion 2.3), 3.0 mL concentrated nitric acid (65 %) (digestion 2.4) or 1.0 mL concentrated hydrochloric acid (36 %) (digestion 2.7). Let cool down and make up to the mark with water. This standard series has Ca concentrations of 0 – 10 – 20 – 200 mg/L.

7.2 Calibration Curve - The emission counts are plot versus mg/L calcium in the standard series.

Remark:

2. A mix standard series can be used for simultaneous measurements with ICP-OES (Al, Ca, Cd, Co, Cr, Cu, Fe, K, Mg, Mn, Na, Ni, P, Pb, S and Zn).

## 8. PROCEDURE

8.1 Measurement - Measure in the standard series, the blanks and sample digests the Ca concentration with the ICP-OES at a wavelength of 317.933 nm. At this wavelength a (fitted) background correction is used.

Remarks:

3. A data management system and system controller are used. In this way the measurements are checked continuously and the data output is in concentration units in the digests.
4. Due to the linearity of the ICP-OES it is possible to calibrate the ICP-OES only with the highest and zero standard of the standard series. As a check all standards can be measured as samples.
5. Beryllium (5 mg/L), at a wavelength of 313.107 nm, is used in the author's laboratory as an internal standard to compensate for matrix effects.

# 9. CALCULATION

9.1 The total calcium content in the dried plant material, expressed in mmol/kg Ca, is calculated by:

$$0.02495 * (a - b) * V / w$$

in which:

a is the concentration of calcium in the sample digest, in mg/L;
b is the concentration of calcium in the blank digest, in mg/L;
V is the total volume of digest at the end of the digestion procedure, in mL;
w is the weight of plant material sample, in g.

# 4.5 DETERMINATION OF CADMIUM

# 4.5.A DETERMINATION OF CADMIUM BY FLAME AAS

## 1. PRINCIPLE OF THE METHOD

1.1 The sample is nebulised into an air-acetylene flame, where it is vaporised; cadmium compounds are atomised and the cadmium atoms thus formed absorb radiation from a hollow cathode lamp. The absorbance is measured at a wavelength of 228.8 nm, using background correction.

1.2 This determination may be carried out on digest 2.3 (microwave digestion with $HNO_3$ - $H_2O_2$ - HF), digest 2.4 (digestion with $HNO_3$ - $H_2O_2$ - HF), digest 2.6 (digestion with $HNO_3$ - $HClO_4$ - $H_2SO_4$) and digest 2.7 (digestion by dry-ashing followed by treatment with HF).

## 2. RANGE AND DETECTION LIMIT

2.1 This procedure yields a standard curve that is linear up to approximately 0.8 mg/L Cd.

2.2 The detection limit is approximately 0.02 mg/L in the digest. The determination limit is approximately 0.06 mg/L (3.0-7.5 mg/kg in dried plant material).

## 3. INTERFERENCES

3.1 To avoid interferences as much as possible, background correction is necessary (deuterium or Smith-Hieftje).

## 4. PRECISION AND ACCURACY

4.1 The reproducibility of determinations by this procedure should give, at reasonable control and thorough sample preparation, a coefficient of variation within 5 % + 5 mg/kg.

## 5. APPARATUS

5.1 Flame atomic absorption spectrophotometer with a device for correcting background absorption.

## 6. REAGENTS

6.1A <u>Stock Solution, Cd concentration 1000 mg/L</u> - Merck nr 1.19777.

6.1B <u>Stock Solution, Cd concentration 1000 mg/L</u> - Dissolve 2.744 g cadmium nitrate tetrahydrate, $Cd(NO_3)_2.4H_2O$, in some water in a 1000-mL volumetric flask and make up to volume.

6.2 <u>Standard Solution, Cd concentration 50 mg/L</u> - Pipette 5.0 mL stock solution (6.1A or 6.1B) into a 100-mL volumetric flask and make up to volume.

<u>Remark</u>:

1. $Cd(NO_3)_2.4H_2O$ may absorb water on standing. The reagent should be standardised by titration with EDTA at pH 10 with Eriochrome Black T as an indicator.

## 7. CALIBRATION AND STANDARDS

7.1 <u>Standard Series</u> - Pipette 0 – 0.10 – 0.20 – 0.40 – 0.80 – 1.20 – 1.60 mL of the standard solution (6.2) into 100-mL volumetric flasks, which already contain 40 mL water. Add either 10 mL concentrated nitric acid (65 %) (digestion 2.3), 3 mL concentrated nitric acid (65 %) (digestion 2.4), 0.45 mL concentrated sulphuric acid (96 %) (digestion 2.6) or 1.0 mL concentrated hydrochloric acid (36 %) (digestion 2.7). Let cool down and make up to the mark with water. This standard series has Cd concentrations of 0 – 0.05 – 0.10 – 0.20 – 0.40 – 0.60 – 0.80 mg/L.

7.2 <u>Calibration Curve</u> - The absorbance (A) is plot versus mg/L cadmium in the standard series.

## 8. PROCEDURE

8.1 <u>Measurement</u> - Measure in the standard series, the blanks and the sample digests the Cd concentration with flame AAS at a wavelength of 228.8 nm, using background correction. Use scale expansion if necessary.

<u>Remark</u>:

2. Samples with high iron content may give too low results when applying a deuterium background correction system (Van der Lee et al., 1987). In this case Smith-Hieftje background correction is necesserary.

## 9. CALCULATION

9.1 The total cadmium content in the dried plant material, expressed in μg/kg Cd, is calculated by:

$1000 * (a - b) * V / w$

in which:

a is the concentration of cadmium in the sample digest, in mg/L;

b is the concentration of cadmium in the blank digest, in mg/L;

V is the total volume of digest at the end of the digestion procedure, in mL;

w is the weight of plant material sample, in g.

## 10. REFERENCES

10.1 Van der Lee, J.J., E.J.M. Temminghoff, V.J.G. Houba and I. Novozamsky. 1987. Background corrections in the determination of Cd and Pb by flame AAS in plant and soil samples with high Fe levels. Appl. Spectrosc. 41: 388-390.

# 4.5.B DETERMINATION OF CADMIUM BY ICP-OES

## 1. PRINCIPLE OF THE METHOD

1.1 Solutions with cadmium compounds are nebulised into an argon plasma, where all components are vaporised. Cadmium compounds are dissociated and excited, and then emit radiation of which the intensity is measured at a wavelength of 214.439 nm.

1.2 This determination may be carried out on digest 2.3 (microwave digestion with $HNO_3$ - $H_2O_2$ - HF), digest 2.4 (digestion with $HNO_3$ - $H_2O_2$ - HF), digest 2.6 (digestion with $HNO_3$ - $HClO_4$ - $H_2SO_4$) and digest 2.7 (digestion by dry-ashing followed by treatment with HF).

## 2. RANGE AND DETECTION LIMIT

2.1 This procedure yields a standard curve that is linear up to at least 2 mg/L Cd.

2.2 The detection limit is approximately 0.002 mg/L in the digest. The determination limit is approximately 0.006 mg/L (0.7 respectively 2.0 mg/kg in dried plant material).

## 3. INTERFERENCES

3.1 No interferences are expected.

## 4. PRECISION AND ACCURACY

4.1 The reproducibility of determinations by this procedure should give, at reasonable control and thorough sample preparation, a coefficient of variation within 5 % + 5 mg/kg.

## 5. APPARATUS

5.1 Inductively coupled plasma optical emission spectrometer.

## 6. REAGENTS

6.1A Stock Solution, Cd concentration 1000 mg/L - Merck nr 1.19777.

6.1B Stock Solution, Cd concentration 1000 mg/L - Dissolve 2.744 g cadmium nitrate tetrahydrate, $Cd(NO_3)_2.4H_2O$, in some water in a 1000-mL volumetric flask and make up to volume.

Remark:

1. $Cd(NO_3)_2.4H_2O$ may absorb water on standing. The reagent should be standardised by titration with EDTA at pH 10 with Eriochrome Black T as an indicator.

6.2 Standard Solution, Cd concentration 100 mg/L - Pipette 0.1 mL concentrated nitric acid (65 % s.p.) in a 100-mL polythene volumetric flask which already contains about 50 mL ultra pure water. Add 10.00 mL stock solution (6.1A or 6.1B) and make up to volume with ultra pure water.

## 7. CALIBRATION AND STANDARDS

7.1 Standard Series - Pipette 0 – 1.00 – 2.00 mL of the standard solution (6.2) into 100-mL volumetric flasks which already contain 40 mL water. Add either 10 mL concentrated nitric acid (65 %) (digestion 2.3), 3.0 mL concentrated nitric acid (65 %) (digestion 2.4), 0.45 mL concentrated sulphuric acid (96 %) (digestion 2.6) or 1.0 mL concentrated hydrochloric acid (36 %) (digestion 2.7). Let cool down and make up to the mark with water. This standard series has Cd concentrations of 0 – 1.0 – 2.0 mg/L.

Remark:

2. A mix standard series can be used for simultaneous measurements with ICP-OES (Al, Ca, Cd, Co, Cr, Cu, Fe, K, Mg, Mn, Na, Ni, P, Pb, S and Zn).

7.2 Calibration Curve - The emission counts are plot versus mg/L cadmium in the standard series.

## 8. PROCEDURE

8.1 Measurement - Measure in the standard series, the blanks and the sample digests the Cd concentration with the ICP-OES at a wavelength of 214.439 nm. At this wavelength a (fitted) background correction is used.

Remarks:

3. A data management system and system controller are used. In this way the measurements are checked continuously and the data output is in concentration units in the digests.
4. Due to the linearity of the ICP-OES it is possible to calibrate the ICP-OES only with the highest and zero standard of the standard series. As a check all standards can be measured as samples.
5. Scandium (5 mg/L), at a wavelength of 255.235 nm, is used in the author's laboratory as an internal standard to compensate for matrix effects.

## 9. CALCULATION

9.1 The total cadmium content in the dried plant material, expressed in mg/kg Cd, is calculated by:

$(a - b) * V / w$

in which:

a is the concentration of cadmium in the sample digest, in mg/L;
b is the concentration of cadmium in the blank digest, in mg/L;
V is the total volume of digest at the end of the digestion procedure, in mL;
w is the weight of plant material sample, in g.

# 4.5.C DETERMINATION OF CADMIUM BY ETA-AAS

## 1. PRINCIPLE OF THE METHOD

1.1 Cadmium ions in the digest are subsequently dried, ashed and vaporised by electrical heating in a graphite furnace. The cadmium atoms thus formed absorb radiation from a hollow cathode lamp. The absorbance is measured at a wavelength of 228.8 nm.

1.2 This determination may be carried out on digest 2.3 (microwave digestion with $HNO_3$ - $H_2O_2$ - HF), digest 2.4 (digestion with $HNO_3$ - $H_2O_2$ - HF), digest 2.6 (digestion with $HNO_3$ - $HClO_4$ - $H_2SO_4$) and digest 2.7 (digestion by dry-ashing followed by treatment with HF).

## 2. RANGE AND DETECTION LIMIT

2.1 This procedure yields a standard curve that is linear up to approximately 4 µg/L Cd.

2.2 The detection limit is approximately 0.05 µg/L in the digest. The determination limit is approximately 0.15 µg/L (7.5-18 µg/kg in the dried plant material).

## 3. INTERFERENCES

3.1 To avoid interferences as much as possible, background correction is necessary (e.g. Zeeman background correction).

## 4. PRECISION AND ACCURACY

4.1 The reproducibility of determinations by this procedure should give, at reasonable control and thorough sample preparation, a coefficient of variation within 10 %.

## 5. APPARATUS

5.1 Electrothermal atomisation (graphite furnace) atomic absorption spectrophotometer with a device for correcting background absorption.

5.2 Polythene cups.

## 6. REAGENTS

6.1A <u>Stock Solution, Cd concentration 1000 mg/L</u> - Merck nr 1.19777.

6.1B Stock Solution, Cd concentration 1000 mg/L - Dissolve 2.744 g cadmium nitrate tetrahydrate, $Cd(NO_3)_2.4H_2O$, in some ultra pure water in a 1000-mL volumetric flask and make up to volume.

6.2 Standard Solution, Cd concentration 10 mg/L - Pipette 1.00 mL stock solution (6.1A or 6.1B) into a 100-mL volumetric flask and make up to volume.

6.3 Diluted Standard Solution, Cd concentration 100 µg/L - Pipette 0.1 mL concentrated nitric acid (65 % s.p.) in a 100-mL polythene volumetric flask which already contains about 50 mL ultra pure water. Add 1.00 mL standard solution (6.2) and make up to volume with ultra pure water.

6.4 Nitric Acid Solution 5 mol/L - Dilute 34.7 mL concentrated nitric acid (65 %) in about 30 mL ultra pure water in a 100-mL volumetric flask. Let cool down and dilute to volume.

6.5 Matrix modifier, Palladium(II)chloride 0.2 % - Dissolve 0.20 g palladium(II) chloride in 0.5 mL concentrated nitric acid (65 %) and heat to dissolve. Heat till almost dry and transfer the solution into a 100-mL volumetric flask and make up to volume.

6.6 Butanol.

6.7 Acidified Triton-X 100 Solution 1 % - Dissolve 1.00 g triton-X in about 20 mL ultra pure water. Transfer the solution into a 100-mL volumetric flask, add 20 mL nitric acid solution (6.4) and make up to volume.

6.8 Propanol-2 Solution 5 % - Dilute 25 mL propanol-2 in some ultra pure water in a 500-mL volumetric flask and make up to volume.

Remark:

1. $Cd(NO_3)_2.4H_2O$ may absorb water on standing. The reagent should be standardised by titration with EDTA at pH 10 with Eriochrome Black T as an indicator.

# 7. CALIBRATION AND STANDARDS

7.1 Standard Series - Pipette 0 – 0.50 – 1.00 – 2.00 – 3.00 – 4.00 mL of the diluted standard solution (6.3) into 100-mL which already contain 40 mL ultra pure water. Add either 10 mL concentrated nitric acid (65 %) (digestion 2.3), 3.0 mL concentrated nitric acid (65 %) (digestion 2.4), 0.45 mL concentrated sulphuric acid (96 %) (digestion 2.6) or 1.0 mL concentrated hydrochloric acid (36 %) (digestion 2.7). Let cool down and make up to the mark with ultra pure water. This standard series has Cd concentrations of 0 – 0.5 – 1 – 2 – 3 – 4 µg/L.

7.2 Calibration Curve - The absorbance (A) is plot versus µg/L cadmium in the standard series.

# 8. PROCEDURE

8.1 Measurement - Pipette 1.00 mL of the standard series, the blanks and sample digests into polythene cups that fit in the automatic sampler of the atomic absorption spectrophotometer. Add 0.05 mL of the acidified triton-X solution (6.7) and mix thoroughly with an electric mini-stirrer. Pipette 1 mL of the matrix modifier (6.5) into another cup, add 0.05 mL of butanol (6.6) and mix. Put standards, modifier, blanks and sample digests in the appropriate place of the sampler. Heat in a graphite furnace according to an appropriate time-temperature programme (see remarks 5 and 6). Measure the absorbance at 228.8 nm in the atomisation phase (use background correction).

Remarks:

2. Every sample should be measured at least three times and for calculation the mean can be used. The relative standard deviation should be less than 2 % for three replicates.
3. Clean all glassware by leaving it for one night in 4 mol/L $HNO_3$. Clean the polythene cups by leaving them for one night in 1 mol/L $HNO_3$; then rinse with ultra pure water and twice with ethanol 96 %. Allow to dry by leaving at room temperature in an inverted position.
4. The wash solution of the automatic sampler contains a 5 % propanol-2 solution (6.8) in order to lower its surface tension and to prevent growth of bacteria.
5. The measurements can be performed with any ETA-AAS system. The present method was worked out using a Varian SpectrAA-300 atomic absorption spectrometer equipped with a graphite tube atomiser, an automatic sampler and a Zeeman-effect background correction system and for a matrix of 0.4 M $HNO_3$ (see remark 7). The operating parameters and temperature programme are given below. Pyrolytically coated partition tubes are used in the author's laboratory. For the given temperature program the use of 0.2 % palladium chloride in 5 % butanol (matrix modifier) is necessary. Butanol (0.05 mL) is added, to 1 mL of the Pd solution, before use, in order to achieve more reproducible drying conditions in the graphite atomiser (Temminghoff, 1990). The sample volume which is injected is 20 μL and of the matrix modifier 5 μL.

| Parameters Cd | Settings |
|---|---|
| lamp current | 5 mA |
| wavelength | 228.8 nm |
| slit width | 0.5 nm |
| measurement mode | peak area |

| Temperature program | | | |
|---|---|---|---|
| Step | Temp (°C) | Time (s) | Sheath gas |
| 1 | 95 | 5.0 | $Ar/H_2$ |
| 2 | 130 | 35.0 | $Ar/H_2$ |
| 3 | 600 | 25.0 | $Ar/H_2$ |
| 4 | 600 | 5.0 | $Ar/H_2$ |
| 5 | 600 | 5.0 | Ar |
| 6 | 600 | 2.5 | - |
| 7 | 2200 | 0.8 | - |
| 8 | 2200 | 3.0 | - |
| 9 | 2600 | 0.2 | Ar |
| 10 | 2600 | 3.0 | Ar |
| Ar = argon, $Ar/H_2$ = 95 % argon and 5 % hydrogen | | | |

The temperature program given here is for the digestion procedure 2.4 ($HNO_3$ - HF - $H_2O_2$). For other digestion procedures the temperature program should be optimised.

6. When using other instruments or matrix modifiers, the optimum temperature values to be set may differ from the values given above. The temperatures mentioned are instrument settings instead of real temperature values; such settings may differ even within two instruments of the same type and should be always checked out experimentally.

## 9. CALCULATION

9.1 The total cadmium content in the dried plant material, expressed in µg/kg Cd, is calculated by:

(a - b) * V / w

in which:

a is the concentration of cadmium in the sample digest, in µg/L;

b is the concentration of cadmium in the blank digest, in µg/L;

V is the total volume of digest at the end of the digestion procedure, in mL;

w is the weight of plant material sample, in g.

## 10. REFERENCES

10.1 Temminghoff, E.J.M. 1990. Signal stabilisation in Electrothermal Atomisation Atomic Absorption Spectrometry by means of addition of butanol. J. Anal. At. Spectrom. 5: 273.

# 4.5.D DETERMINATION OF CADMIUM BY ICP-MS

## 1. PRINCIPLE OF THE METHOD

1.1 Solutions with cadmium compounds are nebulised into an argon plasma, where all components are vaporised. The ions produced are entrained in the plasma gas and introduced into a mass spectrometer, sorted according to their mass-to-charge ratios, and quantified with a channel electron multiplier. Cadmium is determined at mass 114 amu.

1.2 This determination may be carried out on digest 2.3 (microwave digestion with $HNO_3$ - $H_2O_2$ - HF), digest 2.4 (digestion with $HNO_3$ - $H_2O_2$ - HF) and digest 2.7 (digestion by dry-ashing followed by treatment with HF).

## 2. RANGE AND DETECTION LIMIT

2.1 This procedure yields a standard curve that is linear up to at least 50 μg/L Cd.

2.2 The detection limit is approximately 0.003 μg/L in the digest. The determination limit is approximately 0.010 μg/L (1-3 μg/kg in the dried plant material).

## 3. INTERFERENCES

3.1 Since the isotope $^{114}$Sn has the same mass as $^{114}$Cd a correction factor should be used if the solution contains Sn.

## 4. PRECISION AND ACCURACY

4.1 The reproducibility of determinations by this procedure should give, at reasonable control and thorough sample preparation, a coefficient of variation within 10 %.

## 5. APPARATUS

5.1 Inductively coupled plasma mass spectrometer.

## 6. REAGENTS

6.1A <u>Stock Solution, Cd concentration 1000 mg/L</u> - Merck nr 1.02609.

6.1B <u>Stock Solution, Cd concentration 1000 mg/L</u> - Dissolve 2.744 cadmium nitrate tetrahydrate, $Cd(NO_3).4H_2O$, in some ultra pure water in a volumetric flask of 1000 mL. Make up to 1000 mL with ultra pure water.

6.2 Standard Solution, Cd concentration 1 mg/L - Pipette 1 mL concentrated nitric acid (65 % s.p.) in a 1000-mL polythene volumetric flask which already contains about 500 mL ultra pure water. Add 1.00 mL stock solution (6.1A or 6.1B) and make up to volume with ultra pure water.

Remark:

1. $Cd(NO_3).4H_2O$ may absorb water on standing. The reagent should be standardised by titration with EDTA at pH=10 with Eriochrome Black T as an indicator.

## 7. CALIBRATION AND STANDARDS

7.1 Standard Series - pipette 0 – 1.00 – 2.00 – 5.00 mL of the standard solution (6.2) in about 40 mL ultra pure water in a 100-mL polythene volumetric flasks. Add either 1.0 mL concentrated nitric acid (65 %) (digestion 2.3), 0.3 mL concentrated nitric acid (65 %) (digestion 2.4) or 0.1 mL concentrated hydrochloric acid (36 %) (digestion 2.7). Let cool down and make up to the mark with ultra pure water. This standard series has Cd concentrations of 0 – 10 – 20 – 50 μg/L.

7.2 Calibration Curve - The emission counts are plot versus μg/L cadmium in the standard series.

Remark:

2. A mix standard series can be used for simultaneous measurements with ICP-MS (Al, As, Cd, Co, Cr, Cu, Mn, Ni, Pb, Sb, Sn, V and Zn).

## 8. PROCEDURE

8.1 Measurement – Measure in the standard series, the blanks and the sample digests the Cd concentration with the ICP-MS at a mass of 114 amu.

Remarks:

3. A data management system and system controller are used. In this way the measurements are checked continuously and the data output is in concentration units in the digests.
4. Due to the linearity of the ICP-MS it is possible to calibrate the ICP-MS only with the highest and zero standard of the standard series. As a check all standards can be measured as samples.

## 9. CALCULATION

9.1 The total cadmium content in the dried plant material, expressed in μg/kg Cd, is calculated by:

$(a - b) * V / w$

in which:

a is the concentration of cadmium in the sample digest, in μg/L;

b is the concentration of cadmium in the blank digest, in μg/L;

V is the total volume of digest at the end of the digestion procedure, in mL;

w is the weight of plant material sample, in g.

# 4.6 DETERMINATION OF CHLORIDE

# 4.6.A DETERMINATION OF CHLORIDE BY COULOMETRIC TITRATION

## 1. PRINCIPLE OF THE METHOD

1.1 Chloride is titrated with silver ions, which are generated from a silver anode at constant current. The end point is detected by a sudden increase in current through a separate set of silver electrodes, caused by the first excess of free silver ions; this current activates a trigger circuit to stop the pulse counter. The total number of pulses counted is a direct measure of the amount of chloride in the sample.

1.2 This determination may be carried out on extract 3.1 (extraction with water).

Remarks:

1. In a coulometric titration at constant current, the total quantity of electricity passed is derived from the product: current * time. In the present procedure, the apparatus used is calibrated directly in millimoles of chloride.
2. The solution to be measured is acidified to allow a smooth reaction at the reference electrode, and also to prevent adsorption of silver ions on the precipitate and to prevent the precipitation of other silver salts, like phosphate and carbonate.
3. Gelatin is added to the solution to keep the precipitate well suspended. In that way, all newly generated silver ions come instantaneously into contact with the precipitate (see remark 7).

## 2. RANGE AND DETECTION LIMIT

2.1 The detection limit is approximately 0.15 mmol/L in the digest. The determination limit is approximately 0.5 mmol/L (25 mmol/kg in the dried plant material).

## 3. INTERFERENCES

3.1 No interferences are expected if only Cl free filters are used.

## 4. PRECISION AND ACCURACY

4.1 The reproducibility of determinations by this procedure should give, at reasonable control and thorough sample preparation, a coefficient of variation within 10 % + 10 mmol/kg.

## 5. APPARATUS

5.1 Coulometer, fitted for titrations at constant current.

5.2 Working electrode, silver.

5.3 Reference electrode, silver.

5.4 Detection electrodes, silver.

Remark:

4. In the authors' laboratory, a so-called "Chlor-O-Counter" is used; this is a Dutch-made coulometer, which is especially adapted for rapid chloride determinations. The procedure is written for the use of this particular instrument, but in principle any commercially available coulometer will serve the purpose.

## 6. REAGENTS

6.1 Stock Solution, Cl concentration 100 mmol/L - Dissolve 5.844 g of sodium chloride, NaCl (see remark 5) in water in a volumetric flask of 1000 mL and make up to the mark.

6.2 Standard Solution, Cl concentration 10 mmol/L - Pipette 10.0 mL of the stock solution (6.1) into a 100-mL volumetric flask and dilute to volume.

6.3 Acid Mixture - Add 8 mL of concentrated nitric acid (65 %) in about 500 mL water in a 1000-mL volumetric flask. Then add 100 mL of concentrated acetic acid (100 %) and dilute to volume with water and mix.

6.4 Gelatin Solution - Weigh out 600 mg of gelatin (granules of purity grade "album") in a 100-mL beaker. Add 50 mL water and soak for 2 h at room temperature. Thereafter, heat on a water bath or in a beaker with hot water until the gelatin has dissolved. Put into a 100-mL volumetric flask and add 10 mg of thymol (to prevent mould growth) and 10 mg of thymol blue and make up the mark. Store in a refrigerator but do not freeze. Prepare fresh weekly.

Remark:

5. The sodium chloride has to be dried at 200 °C for at least 24 hours just before weighing. Just before drying, any lumps should be cut down so that only fine crystals remain. This is important, because coarse crystals contain much occluded water in their cavities and thus would need a higher drying temperature. The fine crystals should, however, not be pulverised because this enhances deliquescence (= attraction of water) and/or efflorescence (= weathering) later on. After the recommended heating procedure, the sodium chloride should be cooled in a desiccator containing preferably magnesium perchlorate as a desiccant. The dried sodium chloride should be weighed as soon as it has reached the ambient temperature.

## 7. CALIBRATION AND STANDARDS

7.1 Standard Series - No standard series is needed, since the number of pulses (representing the total amount of electricity consumed) is directly proportional to the amount of chloride. Only one standard solution (6.2) must be titrated to establish the actual relation.

# 8. PROCEDURES

8.1 Pretreatment of the Titration Medium - Check whether both electrode couples are scrupulously clean, in particular the detection electrodes; if necessary, clean them with a cleaning powder and wipe with tissue paper. Check also whether the generator electrode is still thick enough (at least 2 mm).

Bring into a tall-shaped 50-mL beaker:

- 20 mL of acid mixture (6.3)
- 1 mL of gelatin solution (6.4)
- 2 mL of standard solution (6.2)

Place the beaker on the titration stand, and immerse the electrodes and the stirrer. Adjust the current at the highest possible value (here: position 80 = 80 x $10^{-9}$ mol $Cl^-$ per pulse), and press the pushbutton "mixing". After a few seconds, press the key "titrating". The titration process can be followed by observing the pulse counter and the formation of precipitate. When the titration is finished, the counter stops automatically. By pressing the key "pipetting", the stirring is stopped and the counter is put to zero again; the number of pulses needs not to be recorded. The titration medium is now ready for the measurements.

8.2 Measurement of Standard Solution - Pipette 1.00 mL of the standard solution (6.2) into the same 50-mL beaker containing the previously titrated solution. Stir for a few seconds and then start the titration at position 10 (= 10 x $10^{-9}$ mol $Cl^-$ per pulse). When the titration is finished, record the number of pulses (about 1000) and put the counter to zero again.

Repeat this procedure two or three times and adjust the current in such a way that 1000 ± 20 pulses are counted.

8.3 Measurement of Samples - Pipette 1.00 mL of the blank or the extract into the same 50-mL beaker, containing the previously titrated solution. Stir for a few seconds and then start the titration at position 5 (= 5 x $10^{-9}$ mol $Cl^-$ per pulse). When the titration is finished, record the number of pulses. Put the counter to zero again, pipette the next sample into the same titration medium, and proceed as before. When the beaker has become completely filled, the procedure has to be started again with a fresh titration medium (see remark 8).

When less than 200 pulses per sample are counted, the measurement must be repeated with more sample (D mL).

Remarks:

6. The silver anode slowly dissolves by forming silver chloride. The consumption of silver is, however, very small: 1 mg for 1 mL of standard solution.
7. The thymol blue is red-coloured below pH 2, so that it indicates whether the titration medium is still acidic enough and also whether the gelatin solution is actually added.
8. The titration medium must contain some precipitate of AgCl, so that during the sample titration the precipitate is formed rapidly enough.
9. When the beaker has become completely filled after several titrations, the titration medium is normally still acidic enough. In that case, it is appropriate to empty the beaker partly and to

proceed with the next samples. In this way, about twenty samples can be handled before the titration medium is "exhausted" (see remark 9).

10. In case the precipitation starts with difficulty, or the counter begins to run irregularly, the titration medium may have become "exhausted", which means that the acid has been consumed by the reaction at the reference electrode. The procedure should then be started again with fresh titration medium.

11. For chloride determinations in plant extracts, the prescribed current of 6 mA is for most samples sufficient. When high concentrations are expected, a higher current value can be chosen.

# 9. CALCULATION

9.1 The chloride content of the dried plant material, expressed in mmol/kg Cl, is calculated by:

0.001 * (a - b) * p / D * V / w

in which:

a is the number of pulses counted in the sample extract;
b is the number of pulses counted in the blank extract;
p is the position of current generator in the Chlor-O-Counter
(p = 5, 10, 20, 40 or 80)
D is the volume of plant extract pipetted, in mL;
V is the volume of extract used for the extraction procedure, in mL;
w is the weight of plant material sample, in g.

Remark:

12. In the calculation formula it is assumed that the standard solution provides 1000 ± 20 pulses. Otherwise, a correction factor 1000/n must be applied, where n = actual number of pulses for the standard solution.

# 4.7 DETERMINATION OF COBALT

# 4.7.A DETERMINATION OF COBALT BY ICP-OES

## 1. PRINCIPLE OF THE METHOD

1.1 Solutions with cobalt compounds are nebulised into an argon plasma, where all components are vaporised. Cobalt compounds are dissociated and excited, and then emit radiation of which the intensity is measured at a wavelength of 231.160 nm.

1.2 This determination may be carried out on digest 2.3 (microwave digestion with $HNO_3$ - $H_2O_2$ - HF), digest 2.4 (digestion with $HNO_3$ - $H_2O_2$ - HF), digest 2.6 (digestion with $HNO_3$ - $HClO_4$ - $H_2SO_4$) and digest 2.7 (digestion by dry-ashing followed by treatment with HF).

## 2. RANGE AND DETECTION LIMIT

2.1 This procedure yields a standard curve that is linear up to at least 2 mg/L Co.

2.2 The detection limit is approximately 0.006 mg/L in the digest. The determination limit is approximately 0.018 mg/L (2.0 respectively 6.0 mg/kg in dried plant material).

## 3. INTERFERENCES

3.1 No interferences are expected.

## 4. PRECISION AND ACCURACY

4.1 The reproducibility of determinations by this procedure should give, at reasonable control and thorough sample preparation, a coefficient of variation within 5 % + 5 mg/kg.

## 5. APPARATUS

5.1 Inductively coupled plasma optical emission spectrometer.

## 6. REAGENTS

6.1 Stock Solution, Co concentration 1000 mg/L - Merck nr 1.02614.

6.2 Standard Solution, Co concentration 100 mg/L - Pipette 0.1 mL concentrated nitric acid (65 % s.p.) in a 100-mL polythene volumetric flask which already contains about 50 mL ultra pure water. Add 10.00 mL stock solution (6.1) and make up to volume with ultra pure water.

## 7. CALIBRATION AND STANDARDS

7.1 Standard Series - Pipette 0 – 1.00 – 2.00 mL of the standard solution (6.2) into 100-mL volumetric flasks which already contain 40 mL water. Add either 10 mL concentrated nitric acid (65 %) (digestion 2.3), 3.0 mL concentrated nitric acid (65 %) (digestion 2.4), 0.45 mL concentrated sulphuric acid (96 %) (digestion 2.6) or 1.0 mL concentrated hydrochloric acid (36 %) (digestion 2.7). Let cool down and make up to the mark with water. This standard series has Co concentrations of 0 – 1.0 – 2.0 mg/L.

Remark:

1. A mix standard series can be used for simultaneous measurements with ICP-OES (Al, Ca, Cd, Co, Cr, Cu, Fe, K, Mg, Mn, Na, Ni, P, S and Zn).

7.2 Calibration Curve - The emission counts are plot versus mg/L cobalt in the standard series.

## 8. PROCEDURE

8.1 Measurement - Measure in the standard series, the blanks and the sample digests the Co concentration with the ICP-OES at a wavelength of 231.160 nm. At this wavelength a (fitted) background correction is used.

Remarks:

3. A data management system and system controller are used. In this way the measurements are checked continuously and the data output is in concentration units in the digests.
4. Due to the linearity of the ICP-OES it is possible to calibrate the ICP-OES only with the highest and zero standard of the standard series. As a check all standards can be measured as samples.
5. Scandium (5 mg/L), at a wavelength of 255.235 nm, is used in the author's laboratory as an internal standard to compensate for matrix effects.

## 9. CALCULATION

9.1 The total cobalt content in the dried plant material, expressed in mg/kg Co, is calculated by:

$(a - b) * V / w$

in which:

a is the concentration of cobalt in the sample digest, in mg/L;

b is the concentration of cobalt in the blank digest, in mg/L;

V is the total volume of digest at the end of the digestion procedure, in mL;

w is the weight of plant material sample, in g.

# 4.7.B DETERMINATION OF COBALT BY ICP-MS

## 1. PRINCIPLE OF THE METHOD

1.1 Solutions with cobalt compounds are nebulised into an argon plasma, where all components are vaporised. The ions produced are entrained in the plasma gas and introduced into a mass spectrometer, sorted according to their mass-to-charge ratios, and quantified with a channel electron multiplier. Cobalt is determined at mass 59 amu.

1.2 This determination may be carried out on digest 2.3 (microwave digestion with $HNO_3$ - $H_2O_2$ - HF), digest 2.4 (digestion with $HNO_3$ - $H_2O_2$ - HF) and digest 2.7 (digestion by dry-ashing followed by treatment with HF).

## 2. RANGE AND DETECTION LIMIT

2.1 This procedure yields a standard curve that is linear up to at least 50 µg/L Co.

2.2 The detection limit is approximately 0.002 µg/L in the digest. The determination limit is approximately 0.006 µg/L (0.7 respectively 2 µg/kg in the dried plant material).

## 3. INTERFERENCES

3.1 Interferences are expected from $^{118}Sn^{++}$, $^{24}Mg^{35}Cl$ and $^{36}Ar^{23}Na$ due to mass overlap with $^{59}Co$.

## 4. PRECISION AND ACCURACY

4.1 The reproducibility of determinations by this procedure should give, at reasonable control and thorough sample preparation, a coefficient of variation within 5 % + 5 µg/kg.

## 5. APPARATUS

5.1 Inductively coupled plasma mass spectrometer.

## 6. REAGENTS

6.1 Stock Solution, Co concentration 1000 mg/L - Merck nr 1.02614.

6.2 Standard Solution, Co concentration 1 mg/L - Pipette 1 mL concentrated nitric acid (65 % s.p.) in a 1000-mL polythene volumetric flask which already contains about

500 mL ultra pure water. Add 1.00 mL stock solution (6.1) and make up to volume with ultra pure water.

# 7. CALIBRATION AND STANDARDS

7.1 Standard Series - pipette 0 – 1.00 – 2.00 – 5.00 mL of the standard solution (6.2) in about 40 mL ultra pure water in a 100-mL polythene volumetric flasks. Add either 1.0 mL concentrated nitric acid (65 %) (digestion 2.3), 0.3 mL concentrated nitric acid (65 %) (digestion 2.4) or 0.1 mL concentrated hydrochloric acid (36 %) (digestion 2.7). Let cool down and make up to the mark with ultra pure water. This standard series has Co concentrations of 0 –10 – 20 – 50 µg/L.

7.2 Calibration Curve - The emission counts are plot versus µg/L cobalt in the standard series.

Remark:

1. A mix standard series can be used for simultaneous measurements with ICP-MS (Al, As, Cd, Co, Cr, Cu, Mn, Ni, Pb, Sb, Sn, V and Zn).

# 8. PROCEDURE

8.1 Measurement – Measure in the standard series, the blanks and the sample digests the Co concentration with the ICP-MS at a mass of 59 amu. Make use of corrections if necessary.

Remarks:

2. A data management system and system controller are used. In this way the measurements are checked continuously and the data output is in concentration units in the digests.
3. Due to the linearity of the ICP-MS it is possible to calibrate the ICP-MS only with the highest and zero standard of the standard series. As a check all standards can be measured as samples.

# 9. CALCULATION

9.1 The total cobalt content in the dried plant material, expressed in µg/kg Co, is calculated by:

$(a - b) * V / w$

in which:

a is the concentration of cobalt in the sample digest, in µg/L;

b is the concentration of cobalt in the blank digest, in µg/L;

V is the total volume of digest at the end of the digestion procedure, in mL;

w is the weight of plant material sample, in g.

# 4.8 DETERMINATION OF CHROMIUM

# 4.8.A DETERMINATION OF CHROMIUM BY ICP-OES

## 1. PRINCIPLE OF THE METHOD

1.1 Solutions with chromium compounds are nebulised into an argon plasma, where all components are vaporised. Chromium compounds are dissociated and excited, and then emit radiation of which the intensity is measured at a wavelength of 205.560 nm.

1.2 This determination may be carried out on digest 2.3 (microwave digestion with $HNO_3$ - $H_2O_2$ - HF), digest 2.4 (digestion with $HNO_3$ - $H_2O_2$ - HF), digest 2.6 (digestion with $HNO_3$ - $HClO_4$ - $H_2SO_4$) and digest 2.7 (digestion by dry-ashing followed by treatment with HF).

## 2. RANGE AND DETECTION LIMIT

2.1 This procedure yields a standard curve that is linear up to at least 2 mg/L Cr.

2.2 The detection limit is approximately 0.003 mg/L in the digest. The determination limit is approximately 0.009 mg/L (0.9 respectively 2.7 mg/kg in dried plant material).

## 3. INTERFERENCES

3.1 No interferences are expected.

## 4. PRECISION AND ACCURACY

4.1 The reproducibility of determinations by this procedure should give, at reasonable control and thorough sample preparation, a coefficient of variation within 10 % + 5 mg/kg.

## 5. APPARATUS

5.1 Inductively coupled plasma optical emission spectrometer.

## 6. REAGENTS

6.1 Stock Solution, Cr concentration 1000 mg/L - Merck nr 1.02613.

6.2 Standard Solution, Cr concentration 100 mg/L - Pipette 0.1 mL concentrated nitric acid (65 % s.p.) in a 100-mL polythene volumetric flask which already contains about 50 mL ultra pure water. Add 10.00 mL stock solution (6.1) and make up to volume with ultra pure water.

## 7. CALIBRATION AND STANDARDS

7.1 Standard Series - Pipette 0 – 1.00 – 2.00 mL of the stock solution (6.1) into 100-mL volumetric flasks which already contain 40 mL water. Add either 10 mL concentrated nitric acid (65 %) (digestion 2.3), 3.0 mL concentrated nitric acid (65 %) (digestion 2.4), 0.45 mL concentrated sulphuric acid (96 %) (digestion 2.6) or 1.0 mL concentrated hydrochloric acid (36 %) (digestion 2.7). Let cool down and make up to the mark with water. This standard series has Cr concentrations of 0 – 1.0 – 2.0 mg/L.

Remark:

1. A mix standard series can be used for simultaneous measurements with ICP-OES (Al, Ca, Cd, Co, Cr, Cu, Fe, K, Mg, Mn, Na, Ni, P, S and Zn).

7.2 Calibration Curve - The emission counts are plot versus mg/L chromium in the standard series.

## 8. PROCEDURE

8.1 Measurement - Measure in the standard series, the blanks and the sample digests the Cr concentration with the ICP-OES at a wavelength of 205.560 nm. At this wavelength a (fitted) background correction is used.

Remarks:

3. A data management system and system controller are used. In this way the measurements are checked continuously and the data output is in concentration units in the digests.
4. Due to the linearity of the ICP-OES it is possible to calibrate the ICP-OES only with the highest and zero standard of the standard series. As a check all standards can be measured as samples.
5. Scandium (5 mg/L), at a wavelength of 255.235 nm, is used in the author's laboratory as an internal standard to compensate for matrix effects.

## 9. CALCULATION

9.1 The total chromium content in the dried plant material, expressed in mg/kg Cr, is calculated by:

$(a - b) * V / w$

in which:

a is the concentration of chromium in the sample digest, in mg/L;

b is the concentration of chromium in the blank digest, in mg/L;

V is the total volume of digest at the end of the digestion procedure, in mL;

w is the weight of plant material sample, in g.

# 4.8.B DETERMINATION OF CHROMIUM BY ICP-MS

## 1. PRINCIPLE OF THE METHOD

1.1 Solutions with chromium compounds are nebulised into an argon plasma, where all components are vaporised. The ions produced are entrained in the plasma gas and introduced into a mass spectrometer, sorted according to their mass-to-charge ratios and quantified with a channel electron multiplier. Chromium is determined at mass 52 or 53 amu.

1.2 This determination may be carried out on digest 2.3 (microwave digestion with $HNO_3$ - $H_2O_2$ - HF), digest 2.4 (digestion with $HNO_3$ - $H_2O_2$ - HF) and digest 2.7 (digestion by dry-ashing followed by treatment with HF).

## 2. RANGE AND DETECTION LIMIT

2.1 This procedure yields a standard curve that is linear up to approximately 50 µg/L Cr.

2.2 The detection limit is approximately 0.01 µg/L in the digests. The determination limit is approximately 0.03 µg/L (0.3 respectively 1 µg/kg in the dried plant material).

## 3. INTERFERENCES

3.1 Interferences can be expected from $^{36}Ar^{16}O$, $^{40}Ar^{12}C$, and $^{35}Cl^{16}O^{1}H$ molecules at mass $^{52}Cr$ and of $^{37}Cl^{16}O$ and $^{36}Ar^{16}O^{1}H$ molecules at mass $^{53}Cr$.

## 4. PRECISION AND ACCURACY

4.1 The reproducibility of determinations by this procedure should give, at reasonable control and thorough sample preparation, a coefficient of variation within 15 %.

## 5. APPARATUS

5.1 Inductively coupled plasma mass spectrometer.

## 6. REAGENTS

6.1 Stock Solution, Cr concentration 1000 mg/L - Merck nr 1.02613.

6.2 Standard Solution, Cr concentration 1 mg/L - Pipette 1 mL concentrated nitric acid (65 % s.p.) in a 1000-mL polythene volumetric flask which already contains about

500 mL ultra pure water. Add 1.00 mL stock solution (6.1) and make up to volume with ultra pure water.

## 7. CALIBRATION AND STANDARDS

7.1 Standard Series - pipette 0 – 1.00 – 2.00 – 5.00 mL of the standard solution (6.2) in about 40 mL ultra pure water in a 100-mL polythene volumetric flasks. Add either 1.0 mL concentrated nitric acid (65 %) (digestion 2.3), 0.3 mL concentrated nitric acid (65 %) (digestion 2.4) or 0.1 mL concentrated hydrochloric acid (36 %) (digestion 2.7). Let cool down and make up to the mark with ultra pure water. This standard series has Cr concentrations of 0 –10 – 20 – 50 µg/L.

7.2 Calibration Curve - The counts per second are plot versus µg/L chromium in the standard series.

Remark:

1. A mix standard series can be used for simultaneous measurements with ICP-MS (Al, As, Cd, Co, Cr, Cu, Mn, Ni, Pb, Sb, Sn, V and Zn).

## 8. PROCEDURE

8.1 Measurement – Measure in the standard series, the blanks and the sample digests the Cr concentration with the ICP-MS at a mass of 52 amu.

Remarks:

2. A data management system and system controller are used. In this way the measurements are checked continuously and the data output is in concentration units in the digest.
3. Due to the linearity of the ICP-MS it is possible to calibrate the ICP-MS only with the highest and zero standard of the standard series. As a check all standards can be measured as samples.

## 9. CALCULATION

9.1 The total chromium content in the dried plant material, expressed in µg/kg Cr, is calculated by:

$(a - b) * V / w$

in which:

a is the concentration of chromium in the sample digest, in µg/L;

b is the concentration of chromium in the blank digest, in µg/L;

V is the total volume of digest at the end of the digestion procedure, in mL;

w is the weight of plant material sample, in g.

# 4.9 DETERMINATION OF COPPER

## 4.9.A DETERMINATION OF COPPER BY FLAME AAS

### 1. PRINCIPLE OF THE METHOD

1.1 The sample is nebulised into an air-acetylene flame, where it is vaporised; copper compounds are atomised and the copper atoms thus formed absorb radiation from a hollow cathode lamp. The absorbance is measured at a wavelength of 324.8 nm.

1.2 This determination may be carried out on digest 2.3 (microwave digestion with $HNO_3$ - $H_2O_2$ - HF), digest 2.4 (digestion with $HNO_3$ - $H_2O_2$ - HF), digest 2.6 (digestion with $HNO_3$ - $HClO_4$ - $H_2SO_4$) and digest 2.7 (digestion by dry-ashing followed by treatment with HF).

### 2. RANGE AND DETECTION LIMIT

2.1 This procedure yields a standard curve that is linear up to approximately 4 mg/L Cu.

2.2 The detection limit is approximately 0.02 mg/L in the digest. The determination limit is approximately 0.06 mg/L (3.0-7.5 mg/kg in the dried plant material).

### 3. INTERFERENCES

3.1 No interferences are expected.

### 4. PRECISION AND ACCURACY

4.1 The reproducibility of determinations by this procedure should give, at reasonable control and thorough sample preparation, a coefficient of variation within 5 % + 5 mg/kg.

### 5. APPARATUS

5.1 Flame atomic absorption spectrophotometer.

### 6. REAGENTS

6.1A Stock Solution, Cu concentration 1000 mg/L - Merck nr 1.19786.

6.1B Stock Solution, Cu concentration 1000 mg/L - Dissolve 3.929 g copper sulphate pentahydrate, $CuSO_4.5H_2O$, in about 500 mL water in a 1000-mL volumetric flask and make up to volume.

6.2 Standard Solution, Cu concentration 100 mg/L - Pipette 10.00 mL stock solution (6.1A or 6.1B) into a 100-mL volumetric flask and make up to volume.

Remark:

1. $CuSO_4.5H_2O$ may lose crystal water on standing. The reagent should be standardised by titration with EDTA at pH 10 with murexide as an indicator.

# 7. CALIBRATION AND STANDARDS

7.1 Standard Series - Pipette 0 – 0.25 – 0.50 – 1.00 – 1.50 – 2.00 mL of the standard solution (6.2) into 100-mL volumetric flasks which already contain 40 mL water. Add either 10 mL concentrated nitric acid (65 %) (digestion 2.3), 3.0 mL concentrated nitric acid (65 %) (digestion 2.4), 0.45 mL concentrated sulphuric acid (96 %) (digestion 2.6) or 1.0 mL concentrated hydrochloric acid (36 %) (digestion 2.7). Let cool down and make up to the mark with water. This standard series has Cu concentrations of 0 – 0.25 – 0.5 – 1.0 – 1.5 – 2.0 mg/L.

7.2 Calibration Curve - The absorbance (A) is plot versus mg/L copper in the standard series.

# 8. PROCEDURE

8.1 Measurement – Measure in the standard series, the blanks and the sample digests the Cu concentration with flame AAS at a wavelength of 324.8 nm, using a just blue (stoichiometric) air-acetylene flame. Use scale expansion if necessary.

# 9. CALCULATION

9.1 The total copper content in dried plant material, expressed in mg/kg Cu, is calculated by:

$(a - b) * V / w$

in which:

a is the concentration of copper in the sample digest, in mg/L;

b is the concentration of copper in the blank digest, in mg/L;

V is the total volume of digest at the end of the digestion procedure, in mL;

w is the weight of plant material sample, in g.

# 4.9.B DETERMINATION OF COPPER BY ICP-OES

## 1. PRINCIPLE OF THE METHOD

1.1 Solutions with copper compounds are nebulised into an argon plasma, where all components are vaporised. Copper compounds are dissociated and excited, and then emit radiation of which the intensity is measured at a wavelength of 327.395 nm.

1.2 This determination may be carried out on digest 2.3 (microwave digestion with $HNO_3$ - $H_2O_2$ - HF), digest 2.4 (digestion with $HNO_3$ - $H_2O_2$ - HF), digest 2.6 (digestion with $HNO_3$ - $HClO_4$ - $H_2SO_4$) and digest 2.7 (digestion by dry-ashing followed by treatment with HF).

## 2. RANGE AND DETECTION LIMIT

2.1 This procedure yields a standard curve that is linear up to at least 2 mg/L Cu.

2.2 The detection limit is approximately 0.003 mg/L in the digest. The determination limit is approximately 0.009 mg/L (1.0 respectively 3.0 mg/kg in the plant material).

## 3. INTERFERENCES

3.1 No interferences are expected.

## 4. PRECISION AND ACCURACY

4.1 The reproducibility of determinations by this procedure should give, at reasonable control and thorough sample preparation, a coefficient of variation within 5 % + 5 mg/kg.

## 5. APPARATUS

5.1 Inductively coupled plasma optical emission spectrometer.

## 6. REAGENTS

6.1A Stock Solution, Cu concentration 1000 mg/L - Merck nr 1.19786.

6.1B Stock Solution, Cu concentration 1000 mg/L - Dissolve 3.929 g copper sulphate pentahydrate, $CuSO_4.5H_2O$, in about 500 mL water in a 1000-mL volumetric flask and make up to volume.

Remark:

1. $CuSO_4.5H_2O$ may lose crystal water on standing. The reagent should be standardised by titration with EDTA at pH 10 with murexide as an indicator.

6.2 Standard Solution, Cu concentration 100 mg/L - Pipette 0.1 mL concentrated nitric acid (65 % s.p.) in a 100-mL polythene volumetric flask which already contains about 50 mL ultra pure water. Add 10.00 mL stock solution (6.1A or 6.1B) and make up to volume with ultra pure water.

## 7. CALIBRATION AND STANDARDS

7.1 Standard Series - Pipette 0 – 1.00 – 2.00 mL of the standard solution (6.2) into 100-mL volumetric flasks which already contain 40 mL water. Add either 10 mL concentrated nitric acid (65 %) (digestion 2.3), 3.0 mL concentrated nitric acid (65 %) (digestion 2.4), 0.45 mL concentrated sulphuric acid (96 %) (digestion 2.6) or 1.0 mL concentrated hydrochloric acid (36 %) (digestion 2.7). Let cool down and make up to the mark with water. This standard series has Cu concentrations of 0 – 1.0 – 2.0 mg/L.

Remark:

2. A mix standard series can be used for simultaneous measurements with ICP-OES (Al, Ca, Cd, Co, Cr, Cu, Fe, K, Mg, Mn, Na, Ni, P, S and Zn).

7.2 Calibration Curve - The emission counts are plot versus mg/L copper in the standard series.

## 8. PROCEDURE

8.1 Measurement - Measure in the standard series, the blanks and the sample digests the Cu concentration with the ICP-OES at a wavelength of 327.395 nm. At this wavelength a (right side) background correction is used.

Remarks:

3. A data management system and system controller are used. In this way the measurements are checked continuously and the data output is in concentration units in the digests.
4. Due to the linearity of the ICP-OES it is possible to calibrate the ICP-OES only with the highest and zero standard of the standard series. As a check all standards can be measured as samples.
5. Beryllium (5 mg/L), at a wavelength of 313.107 nm, is used in the author's laboratory as an internal standard to compensate for matrix effects.

## 9. CALCULATION

9.1 The total copper content in the dried plant material, expressed in mg/kg Cu, is calculated by:

(a - b) * V / w

in which:

a is the concentration of copper in the sample digest, in mg/L;
b is the concentration of copper in the blank digest, in mg/L;
V is the total volume of digest at the end of the digestion procedure, in mL;
w is the weight of plant material sample, in g.

# 4.9.C DETERMINATION OF COPPER BY ETA-AAS

## 1. PRINCIPLE OF THE METHOD

1.1 Copper ions in the digest are subsequently dried, ashed and vaporised by electrical heating in a graphite furnace. The copper atoms thus formed absorb radiation from a hollow cathode lamp. The absorbance is measured at a wavelength of 327.4 nm.

1.2 This determination may be carried out on digest 2.3 (microwave digestion with $HNO_3$ - $H_2O_2$ - HF) digest 2.4 (digestion with $HNO_3$ - $H_2O_2$ - HF), digest 2.6 (digestion with $HNO_3$ - $HClO_4$ - $H_2SO_4$) and digest 2.7 (digestion by dry-ashing followed by treatment with HF).

## 2. RANGE AND DETECTION LIMIT

2.1 This procedure yields a standard curve that is linear up to approximately 50 µg/L Cu.

2.2 The detection limit is approximately 1.2 µg/L in the digest. The determination limit is approximately 3.6 µg/L (0.4 respectively 1.2 mg/kg in the dried plant material).

## 3. INTERFERENCES

3.1 To avoid interferences as much as possible, background correction is necessary (e.g. Zeeman background correction).

Remark:

1. Whereas the flame AAS determination normally uses 324.7 nm as wavelength for measurement, the graphite furnace determination has at 327.4 nm an improved curve linearity.

## 4. PRECISION AND ACCURACY

4.1 The reproducibility of determinations by this procedure should give, at reasonable control and thorough sample preparation, a coefficient of variation within 10 %.

## 5. APPARATUS

5.1 Electrothermal atomisation (graphite furnace) atomic absorption spectrophotometer with a device for correcting background absorption.

5.2 Polythene cups.

# 6. REAGENTS

6.1A Stock Solution, Cu concentration 1000 mg/L - Merck nr 1.19786.

6.1B Stock Solution, Cu concentration 1000 mg/L - Dissolve 3.929 g copper sulphate pentahydrate, $CuSO_4.5H_2O$, in about 500 mL ultra pure water in a 1000-mL volumetric flask and make up to volume.

6.2 Diluted Standard Solution, Cu concentration 1 mg/L - Pipette 1 mL concentrated nitric acid (65 % s.p.) in a 1000-mL polythene volumetric flask which already contains about 500 mL ultra pure water. Add 1.00 mL stock solution (6.1A or 6.1B) and make up to volume with ultra pure water.

6.3 Nitric Acid Solution 5 mol/L - Dilute 34.7 mL concentrated nitric acid (65 %) in about 30 mL ultra pure water in a 100-mL volumetric flask. Let cool down and dilute to volume.

6.4 Matrix modifier, Palladium(II)chloride 0.2 % - Dissolve 0.20 g palladium(II) chloride in 0.5 mL concentrated nitric acid (65 %) and heat to dissolve. Heat till almost dry and transfer the solution into a 100-mL volumetric flask and make up to volume.

6.5 Butanol.

6.6 Acidified Triton-X 100 Solution 1 % - Dissolve 1.00 g triton-X in about 20 mL ultra pure water. Transfer the solution into a 100-mL volumetric flask, add 20 mL nitric acid solution (6.4) and make up to volume.

6.7 Propanol-2 Solution 5 % - Dilute 25 mL propanol-2 in some ultra pure water in a 500-mL volumetric flask and make up to volume.

Remark:

2. $CuSO_4.5H_2O$ may lose crystal water on standing. The reagent should be standardised by titration with EDTA at pH 10 with murexide as an indicator.

# 7. CALIBRATION AND STANDARDS

7.1 Standard Series - Pipette 0 – 1.00 – 2.00 – 3.00 – 4.00 – 5.00 mL of the diluted standard solution (6.2) into 100-mL volumetric flasks which already contain 40 mL of ultra pure water. Add either 10 mL concentrated nitric acid (65 %) (digestion 2.3), 3.0 mL concentrated nitric acid (65 %) (digestion 2.4), 0.45 mL concentrated sulphuric acid (96 %) (digestion 2.6) or 1.0 mL concentrated hydrochloric acid (36 %) (digestion 2.7). Let cool down and make up to the mark with ultra pure water. This standard series has Cu concentrations of 0 – 10 – 20 – 30 – 40 – 50 µg/L.

7.2 Calibration Curve - The absorbance (A) is plot versus mg/L copper in the standard series.

# 8. PROCEDURE

8.1 Measurement - Pipette 1.00 mL of the standard series, the blanks and sample digests into polythene cups that fit in the automatic sampler of the atomic absorption spectrophotometer. Add 0.05 mL of the acidified triton-X solution (6.6) and mix thoroughly with an electric mini-stirrer. Pipette 1 mL of the matrix modifier (6.4) into another cup, add 0.05 mL of butanol (6.5) and mix. Put standards, modifier, blanks and sample digests in the appropriate place of the sampler. Heat in a graphite furnace according to an appropriate time-temperature programme (see remarks 5 and 6). Measure the absorbance at 327.4 nm in the atomisation phase (use background correction).

Remarks:

3. Every sample should be measured at least three times and for calculation the mean can be used. The relative standard deviation should be less than 2 % for three replicates.
4. The wash solution of the automatic sampler contains a 5 % propanol-2 solution (6.7) in order to lower its surface tension and to prevent growth of bacteria.
5. The measurements can be performed with any ETA-AAS system. The present method was worked out using a Varian SpectrAA-300 atomic absorption spectrometer equipped with a graphite tube atomiser, an automatic sampler and a Zeeman-effect background correction system and for a matrix of 0.4 M $HNO_3$ (see remark 7). The operating parameters and temperature programme are given below. Pyrolytically coated partition tubes are used in the author's laboratory. For using the given temperature program 0.2 % palladium chloride in 5 % butanol as matrix modifier is necessary. Butanol (0.05 mL) is added, to 1 mL of the Pd solution, before use, in order to achieve more reproducible drying conditions in the graphite atomiser (Temminghoff, 1990). The sample volume which is injected is 25 μL and of the matrix modifier 5 μL.

| Parameters Cu | Settings |
|---|---|
| Lamp current | 4 mA |
| Wavelength | 327.4 nm |
| slit width | 0.5 nm |
| Measurement mode | Peak area |
| Replicates | 3 |

| Temperature program | | | |
|---|---|---|---|
| Step | Temp (°C) | Time (s) | Sheath gas |
| 1 | 95 | 5.0 | $Ar/H_2$ |
| 2 | 130 | 35.0 | $Ar/H_2$ |
| 3 | 500 | 25.0 | $Ar/H_2$ |
| 4 | 500 | 15.0 | $Ar/H_2$ |
| 5 | 500 | 5.0 | Ar |
| 6 | 500 | 2.5 | - |
| 7 | 2300 | 0.9 | - |
| 8 | 2300 | 3.0 | - |
| 9 | 2500 | 0.3 | Ar |
| 10 | 2500 | 3.0 | Ar |
| 11 | 40 | 17.0 | Ar |
| Ar = argon, $Ar/H_2$ = 95 % argon and 5 % hydrogen | | | |

The temperature program given here is for the digestion procedure 2.4 ($HNO_3$ - HF - $H_2O_2$). For other digestion procedures the temperature program should be optimised.

6. When using other instruments or matrix modifiers, the optimum temperature values to be set may differ from the values given above. The temperatures mentioned are instrument settings instead of real temperature values; such settings may differ even within two instruments of the same type and should be always checked out experimentally.

## 9. CALCULATION

9.1 The total copper content in the dried plant material, expressed in mg/kg Cu, is calculated by:

0.001 * (a - b) * V / w

in which:

a is the concentration of copper in the sample digest, in μg/L;

b is the concentration of copper in the blank digest, in μg/L;

V is the total volume of digest at the end of the digestion procedure, in mL;

w is the weight of plant material sample, in g.

## 10. REFERENCES

10.1 Temminghoff, E.J.M. 1990. Signal stabilisation in Electrothermal Atomisation Atomic Absorption Spectrometry by means of addition of butanol. J. Anal. At. Spectrom. 5: 273.

# 4.9.D DETERMINATION OF COPPER BY ICP-MS

## 1. PRINCIPLE OF THE METHOD

1.1 Solutions with copper compounds are nebulised into an argon plasma, where all components are vaporised. The ions produced are entrained in the plasma gas and introduced into a mass spectrometer, sorted according to their mass-to-charge ratios and quantified with a channel electron multiplier. Copper is determined at mass 63 amu.

1.2 This determination may be carried out on digest 2.3 (microwave digestion with $HNO_3$ - $H_2O_2$ - HF), digest 2.4 (digestion with $HNO_3$ - $H_2O_2$ - HF), digest 2.6 (digestion with $HNO_3$ - $HClO_4$ - $H_2SO_4$) and digest 2.7 (digestion by dry-ashing followed by treatment with HF).

## 2. RANGE AND DETECTION LIMIT

2.1 This procedure yields a standard curve that is linear up to at least 50 µg/L Cu.

2.2 The detection limit is approximately 0.03 µg/L in the digest. The determination limit is approximately 0.09 µg/L (10 respectively 30 µg/kg in the dried plant material).

## 3. INTERFERENCES

3.1 No interferences are expected.

## 4. PRECISION AND ACCURACY

4.1 The reproducibility of determinations by this procedure should give, at reasonable control and thorough sample preparation, a coefficient of variation within 10 %.

## 5. APPARATUS

5.1 Inductively coupled plasma mass spectrometer.

## 6. REAGENTS

6.1A Stock Solution, Cu concentration 1000 mg/L - Merck nr 1.02630.

6.1B Stock Solution, Cu concentration 1000 mg/L - Dissolve 3.929 g copper sulphate pentahydrate, $CuSO_4.5H_2O$, in some ultra pure water in a volumetric flask of 1000 mL. Make up to 1000 mL with ultra pure water.

6.2 Standard Solution, Cu concentration 1 mg/L - Pipette 1 mL concentrated nitric acid (65 % s.p.) in a 1000-mL polythene volumetric flask which already contains about 500 mL ultra pure water. Add 1.00 mL stock solution (6.1A or 6.1B) and make up to volume with ultra pure water.

Remark:

1. $CuSO_4.5H_2O$ may lose crystal water on standing. The reagent should be standardised by titration with EDTA at pH=10 with Murexide as an indicator.

# 7. CALIBRATION AND STANDARDS

7.1 Standard Series - pipette 0 – 1.00 – 2.00 – 5.00 mL of the standard solution (6.2) in about 40 mL ultra pure water in a 100-mL polythene volumetric flasks. Add either 1.0 mL concentrated nitric acid (65 %) (digestion 2.3), 0.3 mL concentrated nitric acid (65 %) (digestion 2.4), 0.045 mL concentrated sulphuric acid (96 %) (digestion 2.6) or 0.1 mL concentrated hydrochloric acid (36 %) (digestion 2.7). Let cool down and make up to the mark with ultra pure water. This standard series has Cu concentrations of 0 –10 – 20 – 50 µg/L.

7.2 Calibration Curve - The counts per second are plot versus µg/L copper in the standard series.

Remark:

2. A mix standard series can be used for simultaneous measurements with ICP-MS (Al, As, Cd, Co, Cr, Cu, Mn, Ni, Pb, Sb, Sn, V and Zn).

# 8. PROCEDURE

8.1 Measurement – Measure in the standard series, the blanks and the sample digests the Cu concentration with the ICP-MS at a mass of 63 amu.

Remarks:

3. A data management system and system controller are used. In this way the measurements are checked continuously and the data output is in concentration units in the digests.
4. Due to the linearity of the ICP-MS it is possible to calibrate the ICP-MS only with the highest and zero standard of the standard series. As a check all standards can be measured as samples.

# 9. CALCULATION

9.1 The total copper content in the dried plant material, expressed in µg/kg Cu, is calculated by:

$(a - b) * V / w$

in which:

a is the concentration of copper in the sample digest, in µg/L;

b is the concentration of copper in the blank digest, in µg/L;

V is the total volume of digest at the end of the digestion procedure, in mL;

w is the weight of plant material sample, in g.

# 4.10 DETERMINATION OF IRON

# 4.10.A DETERMINATION OF IRON BY FLAME AAS

## 1. METHOD OF THE PRINCIPLE

1.1 The sample is nebulised into an air-acetylene flame, where it is vaporised; iron compounds are atomised and the iron atoms thus formed absorb radiation from a hollow-cathode lamp. The absorbance is measured at a wavelength of 248.3 nm.

1.2 This determination may be carried out on digest 2.3 (microwave digestion with $HNO_3$ - $H_2O_2$ - HF), digest 2.4 (digestion with $HNO_3$ - $H_2O_2$ - HF), digest 2.6 (digestion with $HNO_3$ - $HClO_4$ - $H_2SO_4$) and digest 2.7 (digestion by dry-ashing followed by treatment with HF).

## 2. RANGE AND DETECTION LIMIT

2.1 This procedure yields a standard curve, which is linear up to approximately 5 mg/L Fe.

2.2 The detection limit is approximately 0.05 mg/L in the digest. The determination limit is approximately 0.15 mg/L (7.5-18 mg/kg in the dried plant material).

## 3. INTERFERENCES

3.1 No interferences are expected.

## 4. PRECISION AND ACCURACY

4.1 The reproducibility of determinations by this procedure should give, at reasonable control and thorough sample preparation, a coefficient of variation within 5 % + 5 mg/kg.

## 5. APPARATUS

5.1 Flame atomic absorption spectrophotometer.

## 6. REAGENTS

6.1A Stock Solution, Fe concentration 1000 mg/L - Merck nr 1.19781.

6.1B Stock Solution, Fe concentration 1000 mg/L - Dissolve 7.0170 g of ammonium iron sulphate hexahydrate, $(NH_4)_2Fe(SO_4)_2.6H_2O$, in a 1000-mL volumetric flask which contains already 200 mL water and 10 mL concentrated nitric acid (65 %). Make up to the mark with water.

6.2 Standard Solution Fe concentration 50 mg/L - Pipette 25.0 mL of the stock solution (6.1A or 6.1B) into a 500-mL volumetric flask. Add 1 mL concentrated nitric acid (65 %) and make up the mark with water.

Remark:

1. $(NH_4)_2Fe(SO_4)_2.6H_2O$ may lose crystal water on standing. The reagent should be standardised by titration with EDTA at pH 2.5 with sulfosalicylic acid as an indicator.

## 7. CALIBRATION AND STANDARDS

7.1 Standard Series - Pipette 0 – 2.00 – 4.00 – 6.00 – 8.00 – 10.00 mL of the standard solution (6.2) into 100-mL volumetric flasks which already contain 40 mL water. Add either 10 mL concentrated nitric acid (65 %) (digestion 2.3), 3.0 mL concentrated nitric acid (65 %) (digestion 2.4), 0.45 concentrated sulphuric acid (96 %) (digestion 2.6) or 1.0 mL concentrated hydrochloric acid (36 %) (digestion 2.7). Let cool down and make up to the mark with water. This standard series has Fe concentrations of 0 – 1 – 2 – 3 – 4 – 5 mg/L.

7.2 Calibration Curve -The absorbance (A) is plot versus mg/L iron in the standard series.

## 8. PROCEDURE

8.1 Measure in the standard series, the blanks and the sample digests the Fe concentration with flame AAS at a wavelength of 248.3 nm, using a blue (oxidising) air-acetylene flame.

## 9. CALCULATION

9.1 The iron content of the dried plant material, expressed in mg/kg Fe, is calculated by:

(a - b) * V / w

in which:

a is the concentration of iron in the sample digest, in mg/L;

b is the concentration of iron in the blank digest, in mg/L;

V is the total volume of digest at the end of the digestion procedure, in mL;

w is the weight of plant material sample, in g.

# 4.10.B DETERMINATION OF IRON BY ICP-OES

## 1. PRINCIPLE OF THE METHOD

1.1 Solutions with iron compounds are nebulised into an argon plasma, where all components are vaporised. Iron compounds are dissociated and excited, and then emit radiation of which the intensity is measured at a wavelength of 259.94 nm.

1.2 This determination may be carried out on digest 2.3 (microwave digestion with $HNO_3$ - $H_2O_2$ - HF), digest 2.4 (digestion with $HNO_3$ - $H_2O_2$ - HF), digest 2.6 (digestion with $HNO_3$ - $HClO_4$ - $H_2SO_4$) and digest 2.7 (digestion by dry-ashing followed by treatment with HF).

## 2. RANGE AND DETECTION LIMIT

2.1 This procedure yields a standard curve that is linear up to at least 100 mg/L Fe.

2.2 The detection limit is approximately 0.002 mg/L in the digest. The determination limit is approximately 0.06 mg/L (0.66 respectively 2.0 mg/kg in the dried plant material).

## 3. INTERFERENCES

3.1 No interferences are expected.

## 4. PRECISION AND ACCURACY

4.1 The reproducibility of determinations by this procedure should give, at reasonable control and thorough sample preparation, a coefficient of variation within 5 % + 5 mg/kg.

## 5. APPARATUS

5.1 Inductively coupled plasma atomic emission spectrometer.

## 6. REAGENTS

6.1A Stock Solution, Fe concentration 1000 mg/L - Merck nr 1.19781.

6.1B Stock Solution, Fe concentration 1000 mg/L - Dissolve 7.0170 g ammonium iron sulphate hexahydrate, $(NH_4)_2Fe(SO_4)_2.6H_2O$, in a volumetric flask of 1000 mL. Make up to 1000 mL with water.

Remark:

1. $(NH_4)_2Fe(SO_4)_2.6H_2O$ may lose crystal water on standing. The reagent should be standardised by titration with EDTA at pH 2.5 with sulfosalicylic acid as an indicator.

# 7. CALIBRATION AND STANDARDS

7.1 Standard Series - Pipette 0 – 1.00 – 2.00 – 10.00 mL of the standard solution (6.1A or 6.1B) into 100-mL volumetric flasks which already contain 40 mL water. Add either 10 mL concentrated nitric acid (65 %) (digestion 2.3), 3.0 mL concentrated nitric acid (65 %) (digestion 2.4), 0.45 mL concentrated sulphuric acid (96 %) (digestion 2.6) or 1.0 mL concentrated hydrochloric acid (36 %) (digestion 2.7). Let cool down and make up to the mark with water. This standard series has Fe concentrations of 0 – 10 – 20 – 100 mg/L.

Remark:

2. A mix standard series can be used for simultaneous measurements with ICP-OES (Al, Ca, Cd, Co, Cr, Cu, Fe, K, Mg, Mn, Na, Ni, P, S and Zn).

7.2 Calibration Curve - The emission counts are plot versus mg/L iron in the standard series.

# 8. PROCEDURE

8.1 Measurement - Measure in the standard series, the blanks and the sample digests the Fe concentration with the ICP-OES at a wavelength of 238.204 nm. At this wavelength a (fitted) background correction is used.

Remarks:

3. A data management system and system controller are used. In this way the measurements are checked continuously and the data output is in concentration units in the digests.
4. Due to the linearity of the ICP-OES it is possible to calibrate the ICP-OES only with the highest and zero standard of the standard series. As a check all standards can be measured as samples.
5. Scandium (5 mg/L), at a wavelength of 255.235 nm, is used in the author's laboratory as an internal standard to compensate for matrix effects.

# 9. CALCULATION

9.1 The total iron content in the dried plant material, expressed in mg/kg Fe, is calculated by:

$(a - b) * V / w$

in which:

a is the concentration of iron in the sample digest, in mg/L;

b is the concentration of iron in the blank digest, in mg/L;

V is the total volume of digest at the end of the digestion procedure, in mL;

w is the weight of plant material sample, in g.

# 4.11 DETERMINATION OF POTASSIUM

# 4.11.A DETERMINATION OF POTASSIUM BY FLAME AES

## 1. PRINCIPLE OF THE METHOD

1.1 The sample is vaporised in an air-propane flame and the potassium compounds are atomised. The potassium atoms thus formed emit radiation of which the intensity is measured at a wavelength of 766.5 nm.

1.2 This determination may be carried out on digest 2.1 (digestion with $H_2SO_4$ - salicylic acid - $H_2O_2$ - Se), digest 2.2 (digestion with $H_2SO_4$ - salicylic acid - $H_2O_2$), digest 2.3 (microwave digestion with $HNO_3$ - $H_2O_2$ - HF), digest 2.4 (digestion with $HNO_3$ - $H_2O_2$ - HF) and digest 2.7 (digestion by dry-ashing followed by treatment with HF).

## 2. RANGE AND DETECTION LIMIT

2.1 This procedure yields a standard curve that is linear up to approximately 50 mg/L K.

2.2 The detection limit is approximately 2 mg/L in the digest. The determination limit is approximately 6 mg/L (7.7-19 mmol/kg in the dried plant material).

## 3. INTERFERENCES

3.1 To prevent ionisation interferences, cesium is added to act as an ionisation buffer.

## 4. PRECISION AND ACCURACY

4.1 The reproducibility of determinations by this procedure should give, at reasonable control and thorough sample preparation, a coefficient of variation within 5 % + 5 mmol/kg.

## 5. APPARATUS

5.1 Flame atomic emission spectrometer.

## 6. REAGENTS

6.1A Stock Solution, K concentration 5000 mg/L - LPS Benelux 1289.3000.

6.1B Stock Solution, K concentration 5000 mg/L - Dissolve 9.534 g potassium chloride, KCl (see remark 1), in some water in a 1000-mL volumetric flask and make up to the mark with water.

6.2A Cesium Solution, Cs concentration 1.1 g/L - Dissolve 1.4 g cesium chloride, CsCl, in a 1000-mL volumetric flask and make up to the mark with water.

6.2B Cesium-Lanthanum Solution, Cs concentration 1.1 g/L, La concentration 1.1 g/L - Dissolve 1.4 g cesium chloride, CsCl, and 3,43 g lanthanum nitrate hexahydrate, $La(NO_3)_3.6H_2O$, in a 1000-ml volumetric flask and make up to the mark with water.

Remark:

1. The potassium chloride has to be dried at 200 °C for at least 24 hours just before weighing. Just before drying, any lumps should be cut down so that only fine crystals remain. This is important, because coarse crystals contain much occluded water in their cavities and thus would need a higher drying temperature. The fine crystals should, however, not be pulverised because this enhances deliquescence (= attraction of water) and/or efflorescence (= weathering) later on. After the recommended heating procedure, the potassium chloride should be cooled in a desiccator containing preferably magnesium perchlorate as a desiccant. The dried potassium chloride should be weighed as soon as it has reached the ambient temperature.

2. The cesium chloride should be of the highest analytical grade ("pro analysi"), because a lower quality (e.g. "reinst" = "most pure") contains much more Na.

# 7. CALIBRATION AND STANDARDS

7.1 Standard Series - Pipette 0 – 2.00 – 4.00 – 6.00 – 8.00 – 10.00 mL of the standard solution (6.1A or 6.1B) into 100-mL volumetric flasks to about 40 mL water. Add either 4.5 mL concentrated sulphuric acid (96 %) (digestion 2.1 or 2.2), 10 mL concentrated nitric acid (65 %) (digestion 2.3), 3.0 mL concentrated nitric acid (65 %) (digestion 2.4) or 1.0 mL concentrated hydrochloric acid (36 %) (digestion 2.7). Let cool down and make up to the mark with water. The standard series has K concentrations of 0 – 100 – 200 – 300 – 400 – 500 mg/L.

Remark:

3. A mix standard series can be used for simultaneous measurements with Flame-AES (Ca, K and Na). In this case cesium-lanthanum solution (6.2B) is necessary instead of cesium solution (see section 4.4.A or 4.15.A). Lanthanum is used as a releasing agent to release Ca from the interfering compounds.

7.2 Calibration Curve - The emission counts are plot versus mg/L potassium in the standard series.

Remark:

4. The calibration curve should be nearly linear.

# 8. PROCEDURE

8.1 Measurement - Dilute standard series, blanks and sample digests 1+9 (v/v) with the cesium solution or cesium-lanthanum solution (6.2A or 6.2B) (see remark 2) and mix. Measure in the diluted standard series, the diluted blanks and the diluted sample digests the K concentration with flame AES at a wavelength of 766.5 nm, using an air-propane flame.

Remarks:

5. An air-acetylene flame can be used also, since the Cs concentration is high enough to counteract the greater tendency to ionisation.
6. Instead of indicating a wavelength, a simple emission spectrometer ('flame photometer') may only be supplied with a so-called K filter. Make sure that these are interference filters, since glass filters are not selective enough.

# 9. CALCULATION

9.1 The total potassium content in the dried plant material, expressed in mmol/kg K, is calculated by:

0.02558 * (a - b) * V / w

in which:

a is the concentration of potassium in the sample digest, in mg/L;

b is the concentration of potassium in the blank digest, in mg/L;

V is the total volume of digest at the end of the digestion procedure, in mL;

w is the weight of plant material sample, in g.

# 4.11.B DETERMINATION OF POTASSIUM BY ICP-OES

## 1. PRINCIPLE OF THE METHOD

1.1 Solutions with potassium compounds are nebulised into an argon plasma, where all components are vaporised. Potassium compounds are dissociated and excited, and then emit radiation of which the intensity is measured at a wavelength of 766.491 nm.

1.2 This determination may be carried out on digest 2.1 (digestion with $H_2SO_4$ - salicylic acid - $H_2O_2$ - Se), digest 2.2 (digestion with $H_2SO_4$ - salicylic acid - $H_2O_2$), digest 2.3 (microwave digestion with $HNO_3$ - $H_2O_2$ - HF), digest 2.4 (digestion with $HNO_3$ - $H_2O_2$ - HF) and digest 2.7 (digestion by dry-ashing followed by treatment with HF).

## 2. RANGE AND DETECTION LIMIT

2.1 This procedure yields a standard curve that is linear up to at least 20 mg/L K.

2.2 The detection limit is approximately 0.05 mg/L in the digest. The determination limit is approximately 0.14 mg/L (0.4 respectively 1.2 mmol/kg in the dried plant material).

## 3. INTERFERENCES

3.1 No interferences are expected.

## 4. PRECISION AND ACCURACY

4.1 The reproducibility of determinations by this procedure should give, at reasonable control and thorough sample preparation, a coefficient of variation within 5 %.

## 5. APPARATUS

5.1 Inductively coupled plasma optical emission spectrometer.

## 6. REAGENTS

6.1A Stock Solution, K concentration 1000 mg/L - Merck nr 1.19505.

6.1B Stock Solution, K concentration 1000 mg/L - Dissolve 1.9068 g potassium chloride, KCl (see remark 1), in some water in a 1000-mL volumetric flask and make up to the mark with water.

Remark:

1. The potassium chloride has to be dried at 200 °C for at least 24 hours just before weighing. Just before drying, any lumps should be cut down so that only fine crystals remain. This is

important, because coarse crystals contain much occluded water in their cavities and thus would need a higher drying temperature. The fine crystals should, however, not be pulverised because this enhances deliquescence (= attraction of water) and/or efflorescence (= weathering) later on. After the recommended heating procedure, the potassium chloride should be cooled in a desiccator containing preferably magnesium perchlorate as a desiccant. The dried potassium chloride should be weighed as soon as it has reached the ambient temperature.

## 7. CALIBRATION AND STANDARDS

7.1 Standard Series - Pipette 0 – 1.00 – 2.00 mL of the stock solution (6.1A or 6.1B) into 100-mL volumetric flasks which already contain 40 mL water. Add either 4.5 mL concentrated sulphuric acid (96 %) (digestion 2.1 or 2.2), 10 mL concentrated nitric acid (65 %) (digestion 2.3), 3.0 mL concentrated nitric acid (65 %) (digestion 2.4) or 1.0 mL concentrated hydrochloric acid (36 %) (digestion 2.7). Let cool down and make up to the mark with water. This standard series has K concentrations of 0 – 10 – 20 mg/L.

7.2 Calibration Curve - The emission counts are plot versus mg/L potassium in the standard series.

Remark:

2. A mix standard series can be used for simultaneous measurements with ICP-OES (Al, Ca, Cd, Cu, Fe, K, Mg, Mn, Na, Ni, P, S and Zn).

## 8. PROCEDURE

8.1 Measurement - Measure in the standard series, the blanks and the sample digests the K concentration with the ICP-OES at a wavelength of 766.491 nm. At this wavelength a (fitted) background correction is used.

Remarks:

3. A data management system and system controller are used. In this way the measurements are checked continuously and the data output is in concentration units in the digests.
4. Due to the linearity of the ICP-OES it is possible to calibrate the ICP-OES only with the highest and zero standard of the standard series. As a check all standards can be measured as samples.
5. Scandium (5 mg/L) is used in the author's laboratory, at a wavelength of 431.408 nm, as an internal standard to compensate for matrix effects.

## 9. CALCULATION

9.1 The total potassium content in the dried plant material, expressed in mmol/kg K, is calculated by:

$$0.02558 * (a - b) * V / w$$

in which:

a is the concentration of potassium in the sample digest, in mg/L;

b is the concentration of potassium in the blank digest, in mg/L;

V is the total volume of digest at the end of the digestion procedure, in mL;

w is the weight of plant material sample, in g.

# 4.12 DETERMINATION OF MAGNESIUM

# 4.12.A DETERMINATION OF MAGNESIUM BY FLAME AAS

## 1. PRINCIPLE OF THE METHOD

1.1 The sample is vaporised in an air-acetylene flame; magnesium compounds are atomised and the magnesium atoms thus formed absorb radiation from a hollow cathode lamp. The absorbance is measured at a wavelength of 285.2 nm.

1.2 This determination may be carried out on digest 2.1 (digestion with $H_2SO_4$ - salicylic acid - $H_2O_2$ - Se), digest 2.2 (digestion with $H_2SO_4$ - salicylic acid - $H_2O_2$), digest 2.3 (microwave digestion with $HNO_3$ - $H_2O_2$ - HF), digest 2.4 (digestion with $HNO_3$ - $H_2O_2$ - HF) and digest 2.7 (digestion by dry-ashing followed by treatment with HF).

## 2. RANGE AND DETECTION LIMIT

2.1 This procedure yields a standard curve that is linear up to approximately 0.5 mg/L Mg.

2.2 The detection limit is approximately 0.8 mg/L in the digest. The determination limit is approximately 2.4 mg/L (5-16 mmol/kg in the dried plant material).

## 3. INTERFERENCES

3.1 Since Mg atoms are easily captured in the flame into poorly dissociating compounds, in particular with phosphate and aluminium, a releasing agent like lanthanum must be added.

## 4. PRECISION AND ACCURACY

4.1 The reproducibility of determinations by this procedure should give, at reasonable control and thorough sample preparation, a coefficient of variation within 7 % + 5 mmol/kg.

## 5. APPARATUS

5.1 Flame atomic absorption spectrophotometer.

## 6. REAGENTS

6.1A Stock Solution, Mg concentration 1000 mg/L - Merck nr 1.19788.

6.1B Stock Solution, Mg concentration 1000 mg/L - Dissolve 10.130 g magnesium sulphate heptahydrate, $MgSO_4.7H_2O$, in some water in a 1000-mL volumetric flask and make up to the mark with water.

6.2 Standard Solution, Mg concentration 500 mg/L - Pipette 50.0 mL stock solution (6.1A or 6.1B) in a 100-mL volumetric flask and make up to volume with water.

6.3 Lanthanum Solution, La concentration 1 g/L - Dissolve 3.12 g lanthanum nitrate hexahydrate, $La(NO_3)_3.6H_2O$, in some water in a 1000-mL volumetric flask.

Remark:

1. $MgSO_4.7H_2O$ may lose crystal water on standing. The reagent should be standardised by titration with EDTA at pH 10 with Eriochrome Black T as an indicator.

## 7. CALIBRATION AND STANDARDS

7.1 Standard Series - Pipette 0 – 2.00 – 4.00 – 6.00 – 8.00 mL of the standard solution (6.3) into 100-mL volumetric flasks, which contain already 40 mL, water. Add either 4.5 mL concentrated sulphuric acid (96 %) (digestion 2.1 or 2.2), 10 mL concentrated nitric acid (65 %) (digestion 2.3), 3.0 mL concentrated nitric acid (65 %) (digestion 2.4) or 1.0 mL concentrated hydrochloric acid (36 %) (digestion 2.7). Let cool down and make up to the mark with water. This standard series has Mg concentrations of 0 – 10 – 20 – 30 – 40 mg/L.

7.2 Calibration Curve - The absorbance (A) is plot versus mg/L magnesium in the standard series.

Remark:

2. The calibration curve is slightly bent towards the x-axis. In that case calculation by means of linear regression is not allowed.

## 8. PROCEDURE

8.1 Measurement - Dilute the standard series, the blanks and the sample digests 1+39 (v/v) with the lanthanum solution (6.3) and mix. Measure in the diluted standard series, the diluted blanks and the diluted sample digests the Mg concentration with flame AAS at a wavelength of 285.2 nm, using a just blue (stoichiometric) air-acetylene flame.

# 9. CALCULATION

9.1 The total magnesium content of the dried plant material, expressed in mmol/kg Mg, is calculated by:

$0.04114 * (a - b) * V / w$

in which:

a is the concentration of magnesium in the sample digest, in mg/L;

b is the concentration of magnesium in the blank digest, in mg/L;

V is the total volume of digest at the end of the digestion procedure, in mL;

w is the weight of plant material sample, in g.

# 4.12.B DETERMINATION OF MAGNESIUM BY ICP-OES

## 1. PRINCIPLE OF THE METHOD

1.1 Solutions with magnesium compounds are nebulised into an argon plasma, where all components are vaporised. Magnesium compounds are dissociated and excited, and then emit radiation of which the intensity is measured at a wavelength of 280.270 nm.

1.2 This determination may be carried out on digest 2.1 (digestion with $H_2SO_4$ - salicylic acid - $H_2O_2$ - Se), digest 2.2 (digestion with $H_2SO_4$ - salicylic acid - $H_2O_2$), digest 2.3 (microwave digestion with $HNO_3$ - $H_2O_2$ - HF), digest 2.4 (digestion with $HNO_3$ - $H_2O_2$ - HF) and digest 2.7 (digestion by dry-ashing followed by treatment with HF).

## 2. RANGE AND DETECTION LIMIT

2.1 This procedure yields a standard curve that is linear up to at least 40 mg/L Mg.

2.2 The detection limit is approximately 0.001 mg/L in the digest. The determination limit is approximately 0.003 mg/L (0.01 respectively 0.04 mmol/kg in the dried plant material).

## 3. INTERFERENCES

3.1 No interferences are expected.

## 4. PRECISION AND ACCURACY

4.1 The reproducibility of determinations by this procedure should give, at reasonable control and thorough sample preparation, a coefficient of variation within 8 % + 5 mmol/kg.

## 5. APPARATUS

5.1 Inductively coupled plasma optical emission spectrometer.

## 6. REAGENTS

6.1A Stock Solution, Mg concentration 1000 mg/L - Merck nr 1.19788.

6.1B Stock Solution, Mg concentration 1000 mg/L - Dissolve 10.130 g magnesium sulphate heptahydrate, $MgSO_4.7H_2O$, in some water in a 1000-mL volumetric flask and make up to the mark with water.

Remark:

1. $MgSO_4.7H_2O$ may lose crystal water on standing. The reagent should be standardised by titration with EDTA at pH 10 with Eriochrome Black T as an indicator.

## 7. CALIBRATION AND STANDARDS

7.1 Standard Series - Pipette 0 – 1.00 – 2.00 – 4.00 mL of the standard solution (6.1A or 6.1B) into 100-mL volumetric flasks which already contain 40 mL water. Add either 4.5 mL concentrated sulphuric acid (96 %) (digestion 2.1 or 2.2), 10 mL concentrated nitric acid (65 %) (digestion 2.3), 3.0 mL concentrated nitric acid (65 %) (digestion 2.4) or 1.0 mL concentrated hydrochloric acid (36 %) (digestion 2.7). Let cool down and make up to the mark with water. This standard series has Mg concentrations of 0 – 10 – 20 – 40 mg/L.

Remark:

2. A mix standard series can be used for simultaneous measurements with ICP-OES (Al, Ca, Cd, Co, Cr, Cu, Fe, K, Mg, Mn, Na, Ni, P, S and Zn).

7.2 Calibration Curve - The emission counts are plot versus mg/L magnesium in the standard series.

## 8. PROCEDURE

8.1 Measurement - Measure in the standard series, the blanks and the sample digests the Mg concentration with the ICP-OES at a wavelength of 280.270 nm. At this wavelength a (fitted) background correction is used.

Remarks:

3. A data management system and system controller are used. In this way the measurements are checked continuously and the data output is in concentration units in the digests.
4. Due to the linearity of the ICP-OES it is possible to calibrate the ICP-OES only with the highest and zero standard of the standard series. As a check all standards can be measured as samples.
5. Beryllium (5 mg/L), at a wavelength of 313.107 nm, is used in the author's laboratory as an internal standard to compensate for matrix effects.

## 9. CALCULATION

9.1 The total magnesium content in the dried plant material, expressed in mmol/kg Mg, is calculated by:

0.04114 * (a - b) * V / w

in which:

a is the concentration of magnesium in the sample digest, in mg/L;

b is the concentration of magnesium in the blank digest, in mg/L;

V is the total volume of digest at the end of the digestion procedure, in mL;

w is the weight of plant material sample, in g.

# 4.13 DETERMINATION OF MANGANESE

## 4.13.A DETERMINATION OF MANGANESE BY FLAME AAS

### 1. PRINCIPLE OF THE METHOD

1.1 The sample is nebulised into an air-acetylene flame, where it is vaporised; manganese compounds are atomised and the manganese atoms thus formed absorb radiation from a hollow cathode lamp. The absorbance is measured at a wavelength of 279.5 nm.

1.2 This determination may be carried out on digest 2.1 (digestion with $H_2SO_4$ - salicylic acid - $H_2O_2$ - Se), digest 2.2 (digestion with $H_2SO_4$ - salicylic acid - $H_2O_2$), digest 2.3 (microwave digestion with $HNO_3$ - $H_2O_2$ - HF), digest 2.4 (digestion with $HNO_3$ - $H_2O_2$ - HF), digest 2.6 (digestion with $HNO_3$ - $HClO_4$ - $H_2SO_4$) and digest 2.7 (digestion by dry-ashing followed by treatment with HF).

### 2. RANGE AND DETECTION LIMIT

2.1 This procedure yields a standard curve that is linear up to approximately 3 mg/L Mn.

2.2 The detection limit is approximately 0.02 mg/L in the digest. The determination limit is approximately 0.06 mg/L (3-10 mg/kg in the dried plant material).

### 3. INTERFERENCES

3.1 Lanthanum is added to prevent condensed phase interferences.

### 4. PRECISION AND ACCURACY

4.1 The reproducibility of determinations by this procedure should give, at reasonable control and thorough sample preparation, a coefficient of variation within 5 % + 5 mg/kg.

### 5. APPARATUS

5.1 Flame atomic absorption spectrophotometer.

### 6. REAGENTS

6.1A <u>Stock Solution, Mn concentration 1000 mg/L</u> - Merck nr 1.19789.

6.1B <u>Stock Solution, Mn concentration 1000 mg/L</u> - Dissolve 2.877 g potassium permanganate, $KMnO_4$, in about 200 mL water in a beaker and add 1 mL concentrated nitric acid. Reduce the permanganate with a few drops of hydrogen peroxide (30 %) and boil to remove the excess of $H_2O_2$. Transfer the contents of the beaker quantitatively to a 1000-mL volumetric flask and make up to the mark.

6.2 Standard Solution, Mn concentration 100 mg/L - Pipette 10.00 mL stock solution (6.1A or 6.1B) into a 100-mL volumetric flask and make up to the mark with water.

6.3 Lanthanum Solution, La concentration 16.5 g/L - Dissolve 12.86 g lanthanum nitrate hexahydrate, $La(NO_3)_3.6H_2O$, in some water in a 250-mL volumetric flask.

Remark:

1. $KMnO_4$ may decompose on standing by influence of light. The reagent should be standardised, after reduction with hydroxylamine-HCl, by titration with EDTA at pH 10 in the presence of tartrate ions with Eriochrome Black T as an indicator.

# 7. CALIBRATION AND STANDARDS

7.1 Standard Series - Pipette 0 – 1.00 – 2.00 – 3.00 – 4.00 – 5.00 mL of the standard solution (6.2) into 100-mL volumetric flasks to about to about 40 mL water. Add either 4.5 mL concentrated sulphuric acid (96 %) (digestion 2.1 or 2.2), 10 mL concentrated nitric acid (65 %) (digestion 2.3), 3.0 mL concentrated nitric acid (65 %) (digestion 2.4), 0.45 mL concentrated sulphuric acid (96 %) (digestion 2.6) or 1.0 mL concentrated hydrochloric acid (36 %) (digestion 2.7). Let cool down and make up to the mark with water. This standard series has Mn concentrations of 0 – 1 – 2 – 3 – 4 – 5 mg/L.

7.2 Calibration Curve - The absorbance (A) is plot versus mg/L manganese in the standard series.

Remark:

2. The calibration curve is slightly bent towards the x-axis. In that case calculation by means of linear regression is not allowed.

# 8. PROCEDURE

8.1 Measurement - Pipette 5.00 mL of the standard series, the blanks and the sample digests into test tubes. Add 0.50 mL of lanthanum solution (6.3) and mix. Measure in the diluted standard series, the diluted blanks and the diluted sample digest the Mn concentration with flame AAS at a wavelength of 279.5 nm, using a blue (oxidising) air-acetylene flame.

# 9. CALCULATION

9.1 The total manganese content in the dried plant material, expressed in mg/kg Mn, is calculated by:

(a - b) * V / w

in which:

a is the concentration of manganese in the sample digest, in mg/L;
b is the concentration of manganese in the blank digest, in mg/L;
V is the total volume of digest at the end of the digestion procedure, in mL;
w is the weight of plant material sample, in g.

# 4.13.B DETERMINATION OF MANGANESE BY ICP-OES

## 1. PRINCIPLE OF THE METHOD

1.1 Solutions with manganese compounds are nebulised into an argon plasma, where all components are vaporised. Manganese compounds are dissociated and excited, and then emit radiation of which the intensity is measured at a wavelength of 257.610 nm.

1.2 This determination may be carried out on digest 2.1 (digestion with $H_2SO_4$ - salicylic acid - $H_2O_2$ - Se), digest 2.2 (digestion with $H_2SO_4$ - salicylic acid - $H_2O_2$), digest 2.3 (microwave digestion with $HNO_3$ - $H_2O_2$ - HF), digest 2.4 (digestion with $HNO_3$ - $H_2O_2$ - HF), digest 2.6 (digestion with $HNO_3$ - $HClO_4$ - $H_2SO_4$) and digest 2.7 (digestion by dry-ashing followed by treatment with HF).

## 2. RANGE AND DETECTION LIMIT

2.1 This procedure yields a standard curve that is linear up to at least 2 mg/L Mn.

2.2 The detection limit is approximately 0.001 mg/L in the digest. The determination limit is approximately 0.003 mg/L (0.3 respectively 1.0 mg/kg in the dried plant material).

## 3. INTERFERENCES

3.1 No interferences are expected.

## 4. PRECISION AND ACCURACY

4.1 The reproducibility of determinations by this procedure should give, at reasonable control and thorough sample preparation, a coefficient of variation within 5 % + 5 mg/kg.

## 5. APPARATUS

5.1 Inductively coupled plasma optical emission spectrometer.

## 6. REAGENTS

6.1A Stock Solution, Mn concentration 1000 mg/L - Merck nr 1.19789.

6.1B Stock Solution, Mn concentration 1000 mg/L - Dissolve 2.877 g potassium permanganate, $KMnO_4$, in about 200 mL water in a beaker and add 1 mL concentrated nitric acid. Reduce the permanganate with a few drops of hydrogen peroxide (30 %) and boil to remove the excess of $H_2O_2$. Transfer the contents of the beaker quantitatively to a 1000-mL volumetric flask and make up to the mark.

6.2 Standard Solution, Mn concentration 100 mg/L - Pipette 10.00 mL stock solution (6.1A or 6.1B) in a 100-mL volumetric flask and make up to volume with water.

Remark:

1. $KMnO_4$ may decompose on standing by influence of light. The reagent should be standardised, after reduction with hydroxylamine-HCl, by titration with EDTA at pH 10 in the presence of tartrate ions with Eriochrome Black T as an indicator.

# 7. CALIBRATION AND STANDARDS

7.1 Standard Series - Pipette 0 – 1.00 – 2.00 mL of the standard solution (6.2) into 100-mL volumetric flasks which already contain 40 mL water. Add either 4.5 mL concentrated sulphuric acid (96 %) (digestion 2.1 or 2.2), 10 mL concentrated nitric acid (65 %) (digestion 2.3), 3.0 mL concentrated nitric acid (65 %) (digestion 2.4), 0.45 mL concentrated sulphuric acid (96 %) (digestion 2.6) or 1.0 mL concentrated hydrochloric acid (36 %) (digestion 2.7). Let cool down and make up to the mark with water. This standard series has Mn concentrations of 0 – 1.0 – 2.0 mg/L.

7.2 Calibration Curve - The emission counts are plot versus mg/L manganese in the standard series.

Remark:

2. A mix standard series can be used for simultaneous measurements with ICP-OES (Al, Ca, Cd, Co, Cr, Cu, Fe, K, Mg, Mn, Na, Ni, P, S and Zn).

# 8. PROCEDURE

8.1 Measurement - Measure in the standard series, the blanks and the sample digests the Mn concentration with the ICP-OES at a wavelength of 257.610 nm. At this wavelength a (fitted) background correction is used.

Remarks:

3. A data management system and system controller are used. In this way the measurements are checked continuously and the data output is in concentration units in the extracts.
4. Due to the linearity of the ICP-OES it is possible to calibrate the ICP-OES only with the highest and zero standard of the standard series. As a check all standards can be measured as samples.
5. Scandium (5 mg/L), at a wavelength of 255.235 nm, is used in the author's laboratory as an internal standard to compensate for matrix effects.

# 9. CALCULATION

9.1 The total manganese content in the dried plant material, expressed in mg/kg Mn, is calculated by:

$(a - b) * V / w$

in which:

a is the concentration of manganese in the sample digest, in mg/L;

b is the concentration of manganese in the blank digest, in mg/L;

V is the total volume of digest at the end of the digestion procedure, in mL;

w is the weight of plant material sample, in g.

# 4.13.C DETERMINATION OF MANGANESE BY ICP-MS

## 1. PRINCIPLE OF THE METHOD

1.1 Solutions with manganese compounds are nebulised into an argon plasma, where all components are vaporised. The ions produced are entrained in the plasma gas and introduced into a mass spectrometer, sorted according to their mass-to-charge ratios, and quantified with a channel electron multiplier. Manganese is determined at mass 55 amu.

1.2 This determination may be carried out on digest 2.3 (microwave digestion with $HNO_3$ - $H_2O_2$ - HF), digest 2.4 (digestion with $HNO_3$ - $H_2O_2$ - HF), digest 2.6 (digestion with $HNO_3$ - $HClO_4$ - $H_2SO_4$) and digest 2.7 (digestion by dry-ashing followed by treatment with HF).

## 2. RANGE AND DETECTION LIMIT

2.1 This procedure yields a standard curve that is linear up to at least 50 μg/L Mn.

2.2 The detection limit is approximately 0.014 μg/L in the digest. The determination limit is approximately 0.042 μg/L (5 respectively 14 μg/kg in the dried plant material).

## 3. INTERFERENCES

3.1 Interferences are expected from $^{40}Ar^{14}N^{1}H$, $^{39}K^{16}O$ and $^{23}Na^{32}S$ due to mass overlap with $^{55}Mn$.

## 4. PRECISION AND ACCURACY

4.1 The reproducibility of determinations by this procedure should give, at reasonable control and thorough sample preparation, a coefficient of variation within 10 %.

## 5. APPARATUS

5.1 Inductively coupled plasma mass spectrometer.

## 6. REAGENTS

6.1A Stock Solution, Mn concentration 1000 mg/L - Merck nr 1.19789.

6.1B Stock Solution, Mn concentration 1000 mg/L - Dissolve 2.877 g potassium permanganate, $KMnO_4$, in about 200 mL water in a beaker and add 1 mL concentrated nitric acid. Reduce the permanganate with a few drops of hydrogen peroxide (30 %) and boil to remove the excess of $H_2O_2$. Transfer the contents of the beaker quantitatively to a 1000-mL volumetric flask and make up to the mark.

6.2 Standard Solution, Mn concentration 1 mg/L - Pipette 1 mL concentrated nitric acid (65 % s.p.) in a 1000-mL polythene volumetric flask which already contains about 500 mL ultra pure water. Add 1.00 mL stock solution (6.1A or 6.1B) in a 1000-mL volumetric flask and make up to volume with ultra pure water.

Remark:

1. $KMnO_4$ may decompose on standing by influence of light. The reagent should be standardised, after reduction with hydroxylamine-HCl, by titration with EDTA at pH 10 in the presence of tartrate ions with Eriochrome Black T as an indicator.

## 7. CALIBRATION AND STANDARDS

7.1 Standard Series - pipette 0 – 1.00 – 2.00 – 5.00 mL of the standard solution (6.3) in about 40 mL ultra pure water in a 100-mL polythene volumetric flasks. Add either 1.0 mL concentrated nitric acid (65 %) (digestion 2.3), 0.3 mL concentrated nitric acid (65 %) (digestion 2.4), 0.045 mL concentrated sulphuric acid (96 %) (digestion 2.6) or 0.1 mL concentrated hydrochloric acid (36 %) (digestion 2.7). Let cool down and make up to the mark with ultra pure water. This standard series has Mn concentrations of 0 –10 – 20 – 50 µg/L.

7.2 Calibration Curve - The counts per second are plot versus µg/L manganese in the standard series.

Remark:

2. A mix standard series can be used for simultaneous measurements with ICP-MS (Al, As, Cd, Co, Cr, Cu, Mn, Ni, Pb, Sb, Sn, V and Zn).

## 8. PROCEDURE

8.1 Measurement – Measure in the standard series, the blanks and the sample digests the Mn concentration with the ICP-MS at a mass of 55 amu. Make use of corrections if necessary.

Remarks:

3. A data management system and system controller are used. In this way the measurements are checked continuously and the data output is in concentration units in the digests.
4. Due to the linearity of the ICP-MS it is possible to calibrate the ICP-MS only with the highest and zero standard of the standard series. As a check all standards can be measured as samples.

## 9. CALCULATION

9.1 The manganese content of the dried plant material, expressed in µg/kg Mn, is calculated by:

(a - b) * V / w

in which:

a is the concentration of manganese in the sample digest, in μg/L;

b is the concentration of manganese in the blank digest, in μg/L;

V is the total volume of digest at the end of the digestion procedure, in mL;

w is the weight of plant material sample, in g.

# 4.14 DETERMINATION OF NITROGEN FRACTIONS

## 4.14.A DETERMINATION OF TOTAL NITROGEN BY SPECTROPHOTOMETRY

### 1. PRINCIPLE OF THE METHOD

1.1 The determination is based on the Berthelot reaction, in which a phenol derivative (here: salicylate) forms an azo dye in the presence of ammonia and hypochlorite. In alkaline medium, the indophenol thus formed has a green-blue colour, of which the absorbance is measured at a wavelength of 660 nm. This is a measure for the concentration of ammonium, formed by the nitrogen compounds in the sample.

1.2 This determination may be carried out on digest 2.1 (digestion with $H_2SO_4$ - salicylic acid - $H_2O_2$-Se) and digest 2.2 (digestion with $H_2SO_4$ - salicylic acid - $H_2O_2$).

Remarks:

1. To prevent precipitation of hydroxides, EDTA should be added prior to raising the pH.
2. The pH should be adjusted with a buffer solution, since the digests contain variable amounts of acid.
3. Sodium nitroprusside is a catalyst for the Berthelot reaction.

### 2. RANGE AND DETECTION LIMIT

2.1 This procedure yields a standard curve that is linear up to approximately 15 mg/L N.

2.2 The detection limit is approximately 0.1 mg/L in the digest. The determination limit is approximately 0.3 mg/L (1.2 respectively 3.5 mmol/kg in the dried plant material).

### 3. INTERFERENCES

3.1 No interferences are expected.

### 4. PRECISION AND ACCURACY

4.1 The reproducibility of determinations by this procedure should give, at reasonable control and thorough sample preparation, a coefficient of variation within 5 % + 5 mmol/kg.

### 5. APPARATUS

5.1 Spectrophotometer.

# 6. REAGENTS

6.1 Stock Solution, N concentration 2500 mg/L - Dissolve 11.793 g of ammonium sulphate, $(NH_4)_2SO_4$ (see remark 4), in water in a 1000-mL volumetric flask and make up to volume with water.

6.2 Sodium Hydroxide Solution 10 mol/L - Dissolve 200 g of sodium hydroxide, NaOH, in about 400 mL water; allow to cool and make up to 500 mL.

6.3 Salicylate Solution - Dissolve 110 g of salicylic acid, $C_7H_6O_3$, in 105 mL of sodium hydroxide solution (6.2) and make up directly with water to 250 mL. Prepare just before use.

6.4 Buffer Solution pH 12.3 - Dissolve 26.70 g of disodium hydrogen phosphate dihydrate, $Na_2HPO_4.2H_2O$, in some water in a 2-litre volumetric flask. Add 10 mL of the sodium hydroxide solution (6.2) and dilute to volume with water. Measure the pH and adjust if necessary.

6.5 EDTA Solution - Dissolve 4 g of disodium dihydrogen ethylene diamine tetra acetate dihydrate, $Na_2EDTA.2H_2O$, in 100 mL water.

6.6 Hypochlorite Solution - A stock solution, containing approximately 1 M sodium hypochlorite in 0.1 M NaOH, should be purchased commercially. Dilute 20 mL of this stock solution with water to 100 mL. Prepare fresh daily.

6.7 Nitroprusside Solution - Dissolve 50 mg of sodium nitroprusside dihydrate, $Na_2[Fe(CN)_5NO].2H_2O$, in 100 mL water. Prepare just before use.

6.8 Mixed Reagent I - Mix 50 mL of the salicylate solution (6.3) with 100 mL of the nitroprusside solution (6.7) and 5 mL of the EDTA solution (6.5).

6.9 Mixed Reagent II - Mix 200 mL of the buffer solution (6.4) with 50 mL of the hypochlorite solution (6.6).

Remark:

4. The ammonium sulphate has to be dried at 105 °C for at least 2 hours just before weighing. Just before drying, any lumps should be cut down so that only fine crystals remain. This is important, because coarse crystals contain much occluded water in their cavities and thus would need a higher drying temperature. The fine crystals should, however, not be pulverised because this enhances deliquescence (= attraction of water) and/or efflorescence (= weathering) later on. After the recommended heating procedure, the ammonium sulphate should be cooled in a desiccator containing preferably magnesium perchlorate as a desiccant. The dried ammonium sulphate should be weighed as soon as it has reached the ambient temperature.

5. The final solution of sodium hypochlorite should contain 0.7 % ± 0.1 % of active chlorine. This will be true for solutions purchased from a supplier of chemicals; if, however, the hypochlorite was bought as bleach in a supermarket, a check on its concentration is recommended.

# 7. CALIBRATION AND STANDARDS

7.1 Standard Series - Pipette into 100-mL volumetric flasks, that contain already about 40 mL water, 4.5 mL of concentrated sulphuric acid (96 %). Mix and let cool down. Then add 0 – 1.00 – 2.00 – 3.00 – 4.00 – 5.00 – 6.00 mL of the stock solution (6.1) and dilute to volume with water. This standard series has N concentrations of 0 – 25 – 50 – 75 – 100 – 125 – 150 mg/L.

7.2 Calibration Curve - The absorbance (A) is plot versus mg/L nitrogen in the standard series.

# 8. PROCEDURE

8.1 Dilute the standard series, the blanks and sample digests 1 + 9 (v/v) with water. Pipette 0.20 mL of the diluted standard series, the diluted blank digests and the diluted sample digests into test tubes. Add 3.0 mL of mixed reagent I (6.8) and mix. Then add 5.0 mL of mixed reagent II (6.9) and mix. Allow standing for at least 2 h. Measure the absorbance in a 1-cm cuvette at a wavelength of 660 nm. Plot a calibration curve, and read the N concentrations.

Remarks:

6. In 2 hours the blue colour reaches its maximum intensity; it is stable for at least 10 h.
7. The digests obtained according to digestion 2.1 or 2.2 are 0.7 - 0.9 M in $H_2SO_4$. Any further dilutions (if the N concentrations would be higher than the highest standard) should be made with zero standard solution.
8. The test tubes must be used exclusively for this N determination. They must first be cleaned by taking a blank determination; thereafter they should be cleaned only with water.

# 9. CALCULATION

9.1 The nitrogen content of the dried plant material, expressed in mmol/kg N, is calculated by:

0.07139 * (a - b) V / w

in which:

a is the concentration of nitrogen in the sample digest, in mg/L;

b is the concentration of nitrogen in the blank digest, in mg/L;

V is the total volume of digest at the end of the digestion procedure, in mL;

w is the weight of plant material sample, in g.

# 10. REFERENCES

10.1 Novozamsky, I., R. van Eck, J.Ch. van Schouwenburg and I. Walinga. 1974. Total nitrogen determination in plant material by means of the indophenol blue method. Neth.J.agric.Sci. 22: 3-5.

# 4.14.B DETERMINATION OF TOTAL NITROGEN BY SFA

## 1. PRINCIPLE OF THE METHOD

1.1 The determination is based on the Berthelot reaction, in which a phenol derivative (here: salicylate) forms an azo dye in the presence of ammonia and hypochlorite. In alkaline medium, the indophenol thus formed has a green-blue colour, of which the absorbance is measured at a wavelength of 660 nm. This is a measure for the concentration of ammonium, formed by the nitrogen compounds in the sample. The determination is carried out as a so-called segmented-flow analysis (SFA).

1.2 This determination may be carried out on digest 2.1 (digestion with $H_2SO_4$ - salicylic acid - $H_2O_2$ - Se) and digest 2.2 (digestion with $H_2SO_4$ - salicylic acid - $H_2O_2$).

## 2. RANGE AND DETECTION LIMIT

2.1 This procedure yields a standard curve that is linear up to approximately 300 mg/L N.

2.2 The detection limit is approximately 0.1 mg/L in the digest. The determination limit is approximately 0.3 mg/L (1.2 respectively 3.5 mmol/kg in the dried plant material).

## 3. INTERFERENCES

3.1 No interferences are expected.

## 4. PRECISION AND ACCURACY

4.1 The reproducibility of determinations by this procedure should give, at reasonable control and thorough sample preparation, a coefficient of variation within 5 % + 5 mmol/kg.

## 5. APPARATUS

5.1 Segmented-flow analysis system (sampler, pump, nitrogen unit, photometer, computer). In the authors' laboratory, a Skalar Segmented Flow Analyzer is used, but any SFA system will do. Note, that with other systems the flow diagram may require adaptation.

## 6. REAGENTS

6.1 <u>Stock Solution 10000 mg/L N</u> - Dissolve 38.19 g ammonium chloride, $NH_4Cl$ (see remark 1), in a 1000-mL volumetric flask which already contains about 800 mL water and make up to volume.

6.2 Buffer Solution, pH = 5.2 - Dissolve 33 g potassium sodium tartrate, $C_4H_4O_6KNa.4H_2O$, in a 1000-mL volumetric flask, which already contains about 800 mL water. Add 24 g sodium citrate, $C_6H_5O_7Na_3.2H_2O$ and dissolve. Make up to volume and check the pH and correct if necessary with HCl to 5.2 ± 0.1. Add 3 mL Brij 35 (30 %) (Sigma Chemical Co. nr 430AG-6) and mix.

6.3 Sodium Salicylate - Dissolve 25 g sodium hydroxide, NaOH, in a 1000-mL volumetric flask which already contains about 800 mL water. Add 80 g sodium salicylate, $C_7H_5NaO_3$ and make up to volume.

6.4 Sodium Nitroprusside - Dissolve 1 g sodium nitroprusside, $Na_2[Fe(CN)_5NO].2H_2O$, in a 1000-mL volumetric flask which already contains about 800 mL water. Store in a dark coloured bottle. The solution is stable for one week.

6.5 Sodium Dichloroisocyanurate - Dissolve 2 g sodium dichloroisocyanurate, $C_3N_3O_3Cl_2Na.2H_2O$, in a volumetric flask which already contains about 800 mL water and make up to volume and mix. The solution is stable for one week.

6.5 Rinsing Liquid Sampler - Pipette 33 mL concentrated sulphuric acid (96 %) in a volumetric flask, which already contains about 800 mL water. Let cool down and make up to volume.

Remark :

1. The ammonium chloride has to be dried at 200 °C for at least 24 hours just before weighing. Just before drying, any lumps should be cut down so that only fine crystals remain. This is important, because coarse crystals contain much occluded water in their cavities and thus would need a higher drying temperature. The fine crystals should, however, not be pulverised because this enhances deliquescence (= attraction of water) and/or efflorescence (= weathering) later on. After the recommended heating procedure, the ammonium chloride should be cooled in a desiccator containing preferably magnesium perchlorate as a desiccant. The dried ammonium chloride should be weighed as soon as it has reached the ambient temperature.

## 7. CALIBRATION AND STANDARDS

7.1 Standard Series - Pipette into a 100-mL volumetric flasks which already contain 40 mL water 4.5 mL concentrated sulphuric acid (96 %) (digestion 2.1 or 2.2). Let cool down and pipette 0 – 1.00 – 2.00 – 3.00 mL of the stock solution (6.1) and make up to volume with water. The standard series has N concentrations of 0 – 100 – 200 – 300 mg/L.

7.2 Calibration Curve - The absorbance (A) is plot versus mg/L nitrogen in the standard series.

## 8. PROCEDURE

8.1 Measurement - Start the segmented-flow system according to the scheme given in figure 1. Measure the absorbance of the standard series, the blanks and the sample digests at a wavelength of 660 nm.

Remark:

2. A data management system and system controller are used. In this way the measurements are checked continuously and the data output is in concentration units in the digests.

## 9. CALCULATION

9.1 The nitrogen content in the dried plant material, expressed in mmol/kg N, is calculated by:

0.07139 * (a - b) * V / w

in which:

a is the concentration of nitrogen in the sample digest, in mg/L;

b is the concentration of nitrogen in the blank digest, in mg/L;

V is the total volume of digest at the end of the digestion procedure, in mL;

w is the weight of plant material sample, in g.

## 10. REFERENCES

10.1 Krom, M. 1980. Spectrophotometric determination of ammonia; a study of modified Berthelot reaction using salicylate and dichloroisocyanurate. The Analyst April 1980.

10.2 Searle, P.L. 1984. The Berthelot or indophenol reaction and its use in the analysis chemistry of nitrogen. The Analyst 109. April 1984.

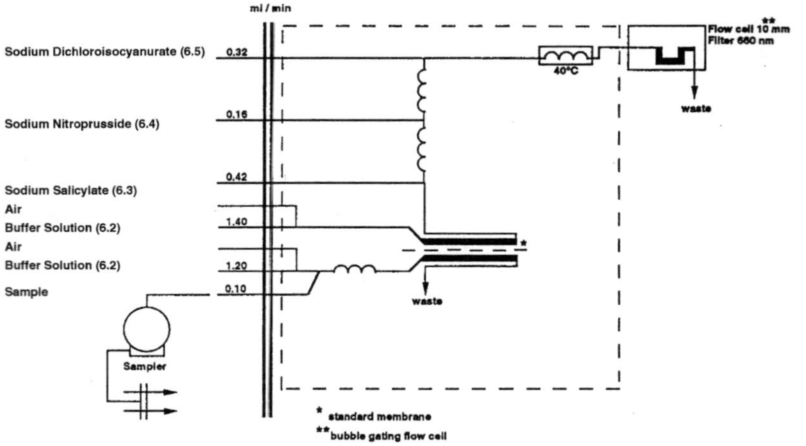


Figure 1 Flow diagram for the determination of nitrogen by SFA.

# 4.14.C DETERMINATION OF NITRATE (+ NITRITE) BY SFA

## 1. PRINCIPLE OF THE METHOD

1.1 In a segmented-flow analysis (SFA) system, the sample is first subjected to dialysis. Nitrate ions from the extract pass the membrane and are taken in an understream of ammonium chloride. The nitrate is then reduced to nitrite by means of copper-coated cadmium. Next, $\alpha$-naphtylamine and sulphanilamide are added, so that in the acid medium present a red-coloured diazo compound is formed. Its absorbance is measured at a wavelength of 540 nm.

1.2 This determination may be carried out on extract 3.1 (extraction with water).

Remarks:

1. The dialysis serves to separate nitrate ions from interfering substances like colloids and coloured organic compounds.
2. The colour reagent is known as the Griess-Ilosvay reagent.
3. Actually, the sum of $NO_3$ and $NO_2$ is determined here.

## 2. RANGE AND DETECTION LIMIT

2.1 This procedure yields a standard curve that is linear up to approximately 65 mg/L N-$NO_3$.

2.2 The detection limit is approximately 0.2 mg/L in the extract. The determination limit is approximately 0.6 mg/L (0.7 respectively 2.0 mmol/kg in the dried plant material).

## 3. INTERFERENCES

3.1 No interferences are expected.

## 4. PRECISION AND ACCURACY

4.1 The reproducibility of determinations by this procedure should give, at reasonable control and thorough sample preparation, a coefficient of variation within 10 %.

## 5. APPARATUS

5.1 Segmented-flow analysis system (sampler, pump, dialysis unit, reduction column, $NO_3$-unit, photometer, computer). In the authors' laboratory, a Skalar Segmented Flow Analyzer is used, but any SFA system will do. Note, that with other systems the flow diagram may require adaptation.

5.2 Reduction column (a U-shaped glass tubing, about 15 cm long and with internal diameter of 2 mm, provided with ferrules for connection to the SFA tubing). It may be purchased filled with copper-coated cadmium from the SFA system manufacturer.

Remark:

4. This SFA arrangement allows the determination of nitrite by simply shortcutting the reduction column.

# 6. REAGENTS

6.1 Stock Solution, N-$NO_3$ concentration 5500 mg/L - Dissolve 39.703 g potassium nitrate, $KNO_3$ (see remark 5), in a 1000-ml flask, which already contains 800 ml water and make up to volume.

6.2 Buffer Solution pH = 8.2 - Dissolve 50 g of ammonium chloride, $NH_4Cl$, in a 1000-mL volumetric flask which contains already about 800 mL water. Adjust the pH to 8.2 with ammonia solution, $NH_4OH$ (25 %) and make up to volume. Add 3 mL of Brij 35 (30 %) (Sigma Chemical Co. nr 430 AG-6) and mix well. Note: it is advised to degas the reagent before adding the Brij 35.

6.3 Diluted Brij 35 - Add 3 mL Brij 35 (30 %) (Sigma Chemical Co. nr 430 AG-6) in a 1000-ml volumetric flask, which already contains about 800 mL water and make up to volume.

6.2 Colour Reagent - Dilute the 150 mL o-phosphoric acid (85 %), $H_3PO_4$, carefully in a volumetric flask, which already contains about 700 mL water. Hereafter add 10 g sulphanilamide, $C_6H_8N_2O_2S$, and 0.5 g $\alpha$-naphthyl ethylene diamine dihydrochloride, $C_{12}H_{16}Cl_2N_2$, and dissolve. Make up to volume and mix well. Store in a dark coloured bottle.

Remark:

5. The potassium nitrate has to be dried at 105 °C for at least 2 hours just before weighing. Just before drying, any lumps should be cut down so that only fine crystals remain. This is important, because coarse crystals contain much occluded water in their cavities and thus would need a higher drying temperature. The fine crystals should, however, not be pulverised because this enhances deliquescence (= attraction of water) and/or efflorescence (= weathering) later on. After the recommended heating procedure, the potassium nitrate should be cooled in a desiccator containing preferably magnesium perchlorate as a desiccant. The dried potassium nitrate should be weighed as soon as it has reached the ambient temperature.

# 7. CALIBRATION AND STANDARDS

7.1 Standard Series - Pipette 0 – 1.00 – 2.00 – 3.00 mL of the stock solution (6.1) into 100-mL volumetric flasks. Make up to the mark with water. This standard series has N-$NO_3$ concentrations of 0 – 22 – 44 – 66 mg/L.

7.2 Calibration Curve - The absorbance (A) is plot versus mg/L $N\text{-}NO_3$ in the standard series.

# 8. PROCEDURE

8.1 Measurement - Start the segmented-flow system according to the scheme given in figure 1. Measure the absorbance of the stardard series, the blanks and the sample digests at a wavelength of 540 nm.

Remark:

6. A data management system and system controller are used. In this way the measurements are checked continuously and the data output is in concentration units in the digests.

# 9. CALCULATION

9.1 The nitrate content in the dried plant material, expressed in mmol/kg $N\text{-}NO_3$, is calculated by:

0.07139 * (a - b) * V / w

in which:

a is the concentration of nitrate in the sample extract, in mg/L;

b is the concentration of nitrate in the blank extract, in mg/L;

V is the volume of water used for the extraction, in mL;

w is the weight of plant material sample, in g.

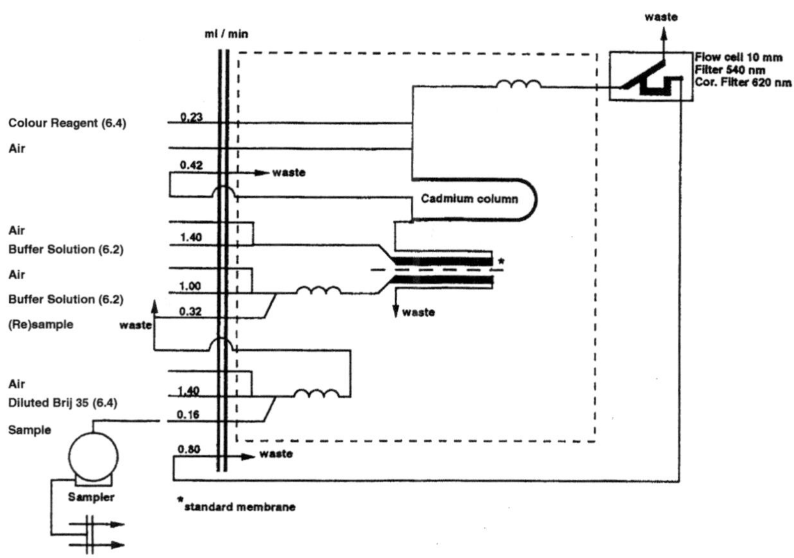


Figure 1 Flow diagram for the determination of $N\text{-}NO_3$ by SFA.

# 4.14.D DETERMINATION OF NITRATE-NITROGEN BY ISE

## 1. PRINCIPLE OF THE METHOD

1.1 After buffering the ionic strength of the water extracts by a phosphate solution, the nitrate ions are detected directly with an ion-selective electrode (ISE).

1.2 This determination may be carried out on extract 3.1 (extraction with water).

Remarks:

1. This method may be used for the determination of nitrate in plant material only when the nitrate content is high (0.5 mmol/kg). At lower levels the interferences by inorganic but especially organic anions are too strong. The nitrate concentration at which the interferences are negligible depends on the plant species (Novozamsky et al., 1983).
2. The addition of phosphate and Al-resin also serves the purpose of buffering the ionic strength.

## 2. RANGE AND DETECTION LIMIT

2.1 The detection limit is approximately 0.1 mmol/L in the extract. The determination limit is approximately 0.3 mmol/L (5 respectively 15 mmol/kg in the dried plant material).

## 3. INTERFERENCES

3.1 Some anionic species interfere: chloride, bicarbonate, phosphate, oxalate, acetate, citrate, succinate, fumarate, malate and malonate. These interferences are partly combated by lowering the pH to 4.0 and addition of cation exchange resin saturated with $Al^{3+}$; the resin lowers the pH further to about 2.5-3. Thus the ionisation is decreased and the organic anions are partly complexed.

## 4. PRECISION AND ACCURACY

4.1 The reproducibility of determinations by this procedure should give, at reasonable control and thorough sample preparation, a coefficient of variation within 15 %.

## 5. APPARATUS

5.1 Potentiometer (pH/mV meter) with which a precision of better than 0.5 mV can be obtained.

5.2 Magnetic stirrer.

5.3 Indicator electrode: nitrate ion-selective electrode.

5.4 Reference electrode: $Hg/Hg_2SO_4$ electrode with saturated $K_2SO_4$ as contact electrolyte.

Remark:

3. In practice, a common calomel electrode can be used, provided that the chloride leakage is very small and no Ag-resin is applied.

# 6. REAGENTS

6.1 Stock solution, $NO_3$ Concentration 20 mmol/L - Dissolve 2.0220 g of potassium nitrate, $KNO_3$ (see remark 4) in some water in a 1000-mL volumetric flask and make up to volume.

6.2 Buffer Solution, pH 4.0 - Dissolve 27.2 g of potassium dihydrogen phosphate, $KH_2PO_4$, in some water and make up to 1 litre.

6.3 Cation Exchange Resin, Dowex 50 W-X 8, 20 - 50 mesh, in $H^+$- or $Na^+$- form.

6.4 Hydrochloric Acid Solution 4 mol/L - Add 330 mL of concentrated hydrochloric acid (36 %) to about 400 mL water and make up to 1 litre.

6.5 Aluminium Solution, Al Concentration 0.50 mol/L - Dissolve 242 g of aluminium chloride hexahydrate, $AlCl_3.6H_2O$, in 2 litre water.

6.6 Silver Solution - Dissolve 0.85 g of silver nitrate, $AgNO_3$, in 100 mL water.

Remark:

4. The potassium nitrate has to be dried at 105 °C for at least 2 hours just before weighing. Just before drying, any lumps should be cut down so that only fine crystals remain. This is important, because coarse crystals contain much occluded water in their cavities and thus would need a higher drying temperature. The fine crystals should, however, not be pulverised because this enhances deliquescence (= attraction of water) and/or efflorescence (= weathering) later on. After the recommended heating procedure, the potassium nitrate should be cooled in a desiccator containing preferably magnesium perchlorate as a desiccant. The dried potassium nitrate should be weighed as soon as it has reached the ambient temperature.

# 7. CALIBRATION AND STANDARDS

7.1 Standard Series - Pipette 0.10 – 0.30 – 1.00 – 3.00 – 10.00 – 30.00 mL of the stock solution (6.1) into 100-mL volumetric flasks and make up to volume. Also the stock solution can be used as a standard solution. This standard series has $NO_3$ concentrations of 0.02 – 0.06 – 0.20 – 0.60 – 2.0 – 6.0 – 20 mmol/L.

Remark:

5. Since there are very large differences in nitrate levels in plant material, a wide standard series is necessary. See also remark 7.

# 8. PROCEDURE

8.1 Preparation of resin - Weigh out 250 g of cation exchange resin (6.3). Introduce a glass wool tuft down into a percolation tube. Transfer the resin to the percolation tube with the help of water to prevent the formation of air bubbles. Leach dropwise with about 1 litre of 4 M hydrochloric acid (6.4). Rinse with water until neutral or weakly acid (pH about 5). Then leach the resin dropwise with approximately 1.5 litre of aluminium solution (6.5) until the pH of the percolate is equal to the pH of the influent (about pH 3). Rinse with water until the leachate is free from chloride; check with silver solution (6.6). Pour the resin slurry on a Büchner funnel on which a filter paper is put. Suck off the water until the resin looks dry; let the resin dry further on the air.

8.2 Measurements - Pipette 5.00 mL of the standard series, the blanks and the sample extracts into 50-mL beakers. Add 5.00 mL of buffer solution (6.2) and a spoonful (0.3 - 0.6 grams) of resin. Place the beaker on a magnetic stirrer and put a small plastic-coated iron bar in the beaker. Immerse both electrodes in the solution, then start stirring. Read the potential (mV) while stirring; take readings at 0.5-min intervals until three subsequent readings have the same value. Plot a calibration curve on semi-logarithmic paper and read the nitrate concentrations.

Remarks:

6. All measurements must be done at constant temperature.
7. If an air bubble sticks on the electrode membrane, then simply take the electrode out of the liquid and immerse it again; this will remove the air bubble.
8. It is recommended to use A3 size graph paper with a 3-decade logarithmic x-axis and a scale unit of 100 mm per decade.

# 9. CALCULATION

9.1 The nitrate content of the dried plant material, expressed in mmol/kg $NO_3$, is calculated by:

a * V / w

in which:

a is the nitrate concentration in the sample extracts, in mmol/L;
V is the volume of extractant used for the extraction of w gram sample, in mL;
w is the weight of plant material sample, in g.

# 10. REFERENCES

10.1 Novozamsky, I., V.J.G. Houba, D. van der Eijk and R. van Eck. 1983. Notes on determinations of nitrate in plant material. Neth.J.Agric.Sci. 31: 239-248.

# 4.14.E DETERMINATION OF NITRITE BY SFA

## 1. PRINCIPLE OF THE METHOD

1.1 In a segmented-flow analysis (SFA) system, the sample is first subjected to dialysis. Nitrite ions from the extract pass the membrane and are taken in an understream of ammonium chloride. By addition of $\alpha$-naphtylamine and sulphanilamide in acid medium, the nitrite forms a red-coloured diazo compound, of which the absorbance is measured at a wavelength of 540 nm.

Remarks:
1. The dialysis serves to separate ions from interfering substances like colloids and coloured organic compounds.
2. The colour reagent is known as the Griess-Ilosvay reagent.

1.2 This determination may be carried out on extract 3.1 (extraction with water).

## 2. RANGE AND DETECTION LIMIT

2.1 This procedure yields a standard curve that is linear up to approximately 3 mg/L N-$NO_2$.

2.2 The detection limit is approximately 0.1 mg/L in the extract. The determination limit is approximately 0.3 mg/L (0.3 respectively 1 mmol/kg in the dried plant material).

## 3. INTERFERENCES

3.1 No interferences are expected.

## 4. PRECISION AND ACCURACY

4.1 The reproducibility of determinations by this procedure should give, at reasonable control and thorough sample preparation, a coefficient of variation within 10 mmol + 10%.

## 5. APPARATUS

5.1 Segmented-flow analysis system (sampler, pump, dialysis unit, $NO_2$-unit, computer). In the authors' laboratory, a Skalar Segmented Flow Analyzer is used, but any SFA system will do. Note that with other systems the flow diagram may require adaptation.

Remark:

3. In the authors' laboratory the SFA arrangement of determination 4.14.C (nitrate + nitrite) is used by simply shortcutting the Cu-Cd reduction column.

# 6. REAGENTS

6.1 Stock Solution, N-$NO_2$ concentration 2000 mg/L - Dissolve 9.852 g sodium nitrite, $NaNO_2$ (see remark 4), in a 1000-ml flask, which already contains 800 ml water and make up to volume.

6.2 Standard Solution, N-$NO_2$ concentration 100 mg/L - Pipette 5 mL stock solution (6.1) in a 100-mlL volumetric flask and make up to volume with water.

6.3 Buffer Solution, pH = 8.2 - Dissolve 50 g of ammonium chloride, $NH_4Cl$, in a 1000-mL volumetric flask which contains already about 800 mL water. Adjust the pH to 8.2 with ammonia solution, $NH_4OH$ (25 %) and make up to volume. Add 3 mL of Brij 35 (30 %) (Sigma Chemical Co. nr 430AG-06) and mix well. Note: it is advised to degas the reagent before adding the Brij 35.

6.4 Diluted Brij 35 - Add 3 mL Brij 35 (30 %) (Sigma Chemical Co. nr 430AG-06) in a volumetric flask which contains already contains about 800 mL water and make up to volume.

6.3 Colour Reagent - Dilute the 150 mL o-phosphoric acid (85 %), $H_3PO_4$, carefully in a volumetric flask, which already contains about 700 mL water. Hereafter add 10 g sulphanilamide, $C_6H_8N2O_2S$, and 0.5 g α-naphthyl ethylene diamine dihydrochloride, $C_{12}H_{16}Cl_2N_2$, and dissolve. Make up to volume and mix well. Store in a dark coloured bottle.

Remark:

4. The sodium nitrite has to be dried at 105 °C for at least 2 hours just before weighing. Just before drying, any lumps should be cut down so that only fine crystals remain. This is important, because coarse crystals contain much occluded water in their cavities and thus would need a higher drying temperature. The fine crystals should, however, not be pulverised because this enhances deliquescence (= attraction of water) and/or efflorescence (= weathering) later on. After the recommended heating procedure, the sodium nitrite should be cooled in a desiccator containing preferably magnesium perchlorate as a desiccant. The dried sodium nitrite should be weighed as soon as it has reached the ambient temperature.

# 7. CALIBRATION AND STANDARDS

7.1 Standard Series - Pipette into 100-mL volumetric flasks 0 – 1.00 – 2.00 – 3.00 mL of the standard solution (6.6) and make up to volume. This standard series has $NO_2$ concentrations of 0 – 10 – 20 – 30 mg/L.

7.2 Calibration Curve - The absorbance (A) is plot versus mg/L N-$NO_2$ in the standard series.

# 8. PROCEDURE

8.1 Measurement - Transfer blanks and sample extracts into polycarbonate test tubes, place them in the automatic sampler and put the segmented-flow system into operation according to the scheme given in figure 1. Measure the absorbance at a wavelength of 540 nm.

Remark:

5. A data management system and system controller are used. In this way the measurements are checked continuously and the data output is in concentration units in the digests.

# 9. CALCULATION

9.1 The nitrite content in the dried plant material, expressed in mmol/kg N-$NO_2$, is calculated by:

0.07139 * (a - b) * V / w

in which:

a is the concentration of nitrite-nitrogen in the sample extract, in mg/L;
b is the concentration of nitrite-nitrogen in the blank extract, in mg/L;
V is the total volume of water used for the extraction, in mL;
w is the weight of plant material sample, in g.

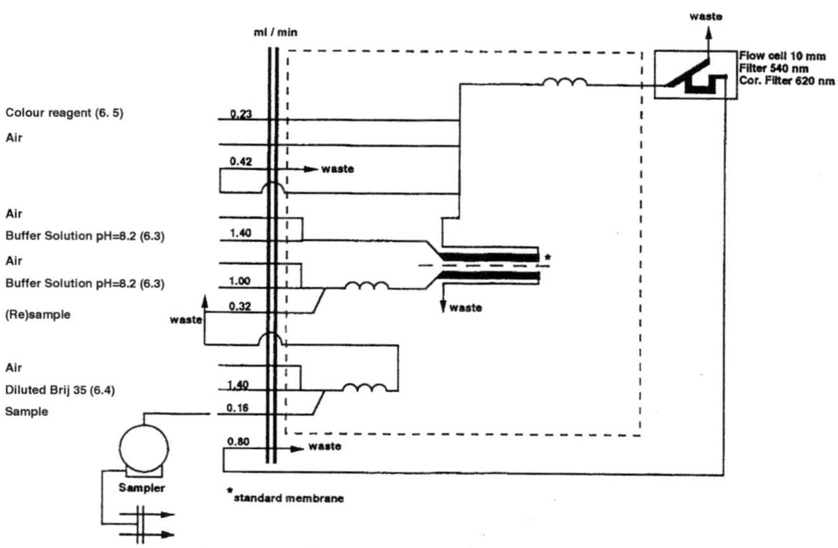


Figure 1. Flow diagram for the determination ot N-$NO_2$ by SFA.

# 4.15 DETERMINATION OF SODIUM

# 4.15.A DETERMINATION OF SODIUM BY FLAME AES

## 1. PRINCIPLE OF THE METHOD

1.1 The sample is vaporised in an air-propane flame and the sodium compounds are atomised. The sodium atoms thus formed emit radiation of which the intensity is measured at a wavelength of 589.0 nm.

1.2 This determination may be carried out on 2.1 digest (digestion with $H_2SO_4$ - salicylic acid - $H_2O_2$ - Se), digest 2.2 (digestion with $H_2SO_4$ - salicylic acid - $H_2O_2$), digest 2.3 (microwave digestion with $HNO_3$ - $H_2O_2$ - HF), digest 2.4 (digestion with $HNO_3$ - $H_2O_2$ - HF) and digest 2.7 (digestion by dry-ashing followed by treatment with HF).

## 2. RANGE AND DETECTION LIMIT

2.1 This procedure yields a standard curve that is linear up to approximately 15 mg/L Na.

2.2 The detection limit is approximately 1.0 mg/L in the digest. The determination limit is approximately 3.0 mg/L (6.0-21 mmol/kg in the dried plant material).

## 3. INTERFERENCES

3.1 To prevent ionisation interferences, cesium is added to act as an ionisation buffer.

3.2 Since the sodium atomic emission line finds itself in the wavelength region of a calcium oxide emission band, too high results will be found if calcium is present in the digest. Lanthanum is used as a releasing agent to release Ca from the interfering compounds.

## 4. PRECISION AND ACCURACY

4.1 The reproducibility of determinations by this procedure should give, at reasonable control and thorough sample preparation, a coefficient of variation within 10 %.

## 5. APPARATUS

5.1 Flame atomic emission spectrometer.

# 6. REAGENTS

6.1A Stock Solution, Na concentration 1000 mg/L - Merck nr 1.19507.

6.1B Stock Solution, Na concentration 1000 mg/L - Dissolve 2.542 g sodium chloride, NaCl (see remark 1), in some water in a 1000-mL volumetric flask and make up to the mark with water.

6.2A Cesium Solution, Cs concentration 1.1 g/L - Dissolve 1.4 g cesium chloride, CsCl, in a 1000-mL volumetric flask and make up to the mark with water.

6.2B Cesium-Lanthanum Solution, Cs concentration 1.1 g/L, La concentration 1.1 g/L - Dissolve 1.4 g cesium chloride, CsCl, and 3.43 g lanthanum nitrate hexahydrate, $La(NO_3)_3.6H_2O$, in a 1000-mL volumetric flask and make up to the mark with water.

Remarks:

1. The sodium chloride has to be dried at 200 °C for at least 24 hours just before weighing. Just before drying, any lumps should be cut down so that only fine crystals remain. This is important, because coarse crystals contain much occluded water in their cavities and thus would need a higher drying temperature. The fine crystals should, however, not be pulverised because this enhances deliquescence (= attraction of water) and/or efflorescence (= weathering) later on. After the recommended heating procedure, the sodium chloride should be cooled in a desiccator containing preferably magnesium perchlorate as a desiccant. The dried sodium chloride should be weighed as soon as it has reached the ambient temperature.
2. The cesium chloride should be of the highest analytical grade ('pro analysis'), because a lower quality (e.g. 'reinst' = 'most pure') contains much more Na.
3. The cesium solution (6.2A) should be used for digests 2.1 and 2.2, whereas the cesium-lanthanum solution (6.2B) should be used for digests 2.3, 2.4 and 2.7 (see Interferences 3.2).

# 7. CALIBRATION AND STANDARDS

7.1 Standard Series - Pipette 0 – 3.00 – 6.00 – 9.00 – 12.00 – 15.00 mL of the stock solution (6.1A or 6.1B) into 100-mL volumetric flasks which already contain 40 mL water. Add 4.5 mL concentrated sulphuric acid (96 %) (digestion 2.1 or 2.2), 10 mL concentrated nitric acid (65 %) (digestion 2.3), 3.0 mL concentrated nitric acid (65 %) (digestion 2.4) or 1.0 mL concentrated hydrochloric acid (36 %) (digestion 2.7). Let cool down and make up to the mark with water.
This standard series has Na concentrations of 0 – 30 – 60 – 90 – 120 – 150 mg/L.

7.2 Calibration Curve - The emission counts are plot versus mg/L sodium in the standard series.

Remarks:

4. A mix standard series can be used for simultaneous measurements with Flame-AES (Ca, K and Na).
5. The calibration curve should be nearly linear.

# 8. PROCEDURE

8.1 Measurement - Dilute the standard series, the blanks and the sample digests 1 + 9 (v/v) either with the cesium solution (6.2A) (digests 2.1 and 2.2) or with the cesium-lanthanum solution (6.2B) (digests 2.3, 2.4 and 2.7). Measure in the diluted standard series, the blanks and the sample digests the Na concentration with flame AES at a wavelength of 598.0 nm, using an air-propane flame.

Remarks:

6. An air-acetylene flame can be used also, since the Cs concentration is high enough to counteract the greater tendency to ionisation.
7. Instead of indicating a wavelength, a simple emission spectrometer ("flame photometer") may only be supplied with a so-called Na filter. Make sure that these are interference filters, since glass filters are not selective enough.

# 9. CALCULATION

9.1 The total sodium content in the dried plant material, expressed in mmol/kg Na, is calculated by:

$0.04350 * (a - b) * V / w$

in which:

a is the concentration of sodium in the sample digest, in mg/L;

b is the concentration of sodium in the blank digest, in mg/L;

V is the total volume of digest at the end of the digestion procedure, in mL;

w is the weight of plant material sample, in g.

# 4.15.B DETERMINATION OF SODIUM BY ICP-OES

## 1. PRINCIPLE OF THE METHOD

1.1 Solutions with sodium compounds are nebulised into an argon plasma, where all components are vaporised. Sodium compounds are dissociated and excited, and then emit radiation of which the intensity is measured at a wavelength of 589.592 nm.

1.2 This determination may be carried out on 2.1 digest (digestion with $H_2SO_4$ - salicylic acid - $H_2O_2$ - Se), digest 2.2 (digestion with $H_2SO_4$ - salicylic acid - $H_2O_2$), digest 2.3 (microwave digestion with $HNO_3$ - $H_2O_2$ - HF), digest 2.4 (digestion with $HNO_3$ - $H_2O_2$ - HF) and digest 2.7 (digestion by dry-ashing followed by treatment with HF).

## 2. RANGE AND DETECTION LIMIT

2.1 This procedure yields a standard curve that is linear up to at least 60 mg/L Na.

2.2 The detection limit is approximately 0.005 mg/L in the digest. The determination limit is approximately 0.015 mg/L (0.07 respectively 0.22 mmol/kg in the dried plant material).

## 3. INTERFERENCES

3.1 Interferences are expected from Ba (589.612 nm).

## 4. PRECISION AND ACCURACY

4.1 The reproducibility of determinations by this procedure should give, at reasonable control and thorough sample preparation, a coefficient of variation within 5 %.

## 5. APPARATUS

5.1 Inductively coupled plasma optical emission spectrometer.

## 6. REAGENTS

6.1A <u>Stock Solution, Na concentration 1000 mg/L</u> - Merck nr 1.19507.

6.1B <u>Stock Solution, Na concentration 1000 mg/L</u> - Dissolve 2.542 g sodium chloride, NaCl (see remark 1), in some water in a 1000-mL volumetric flask and make up to the mark with water.

<u>Remark</u>:

1. The sodium chloride has to be dried at 200 °C for at least 24 hours just before weighing. Just before drying, any lumps should be cut down so that only fine crystals remain. This is

important, because coarse crystals contain much occluded water in their cavities and thus would need a higher drying temperature. The fine crystals should, however, not be pulverised because this enhances deliquescence (= attraction of water) and/or efflorescence (= weathering) later on. After the recommended heating procedure, the sodium chloride should be cooled in a desiccator containing preferably magnesium perchlorate as a desiccant. The dried sodium chloride should be weighed as soon as it has reached the ambient temperature.

## 7. CALIBRATION AND STANDARDS

7.1 Standard Series - Pipette 0 – 1.00 – 2.00 – 6.00 mL of the stock solution (6.1A or 6.1B) into 100-mL volumetric flasks, which already contain 40 mL water. Add either 4.5 mL concentrated sulphuric acid (96 %) (digestion 2.1 or 2.2), 10 mL concentrated nitric acid (65 %) (digestion 2.3), 3.0 mL concentrated nitric acid (65 %) (digestion 2.4) or 1.0 mL concentrated hydrochloric acid (36 %) (digestion 2.7). Let cool down and make up to the mark with water. This standard series has Na concentrations of 0 – 10 – 20 – 60 mg/L.

7.2 Calibration Curve - The emission counts are plot versus mg/L sodium in the standard series.

Remark:

2. A mix standard series can be used for simultaneous measurements with ICP-OES (Al, Ca, Cd, Co, Cr, Cu, Fe, K, Mg, Mn, Na, Ni, P, S and Zn).

## 8. PROCEDURE

8.1 Measurement - Measure in the standard series, the blanks and the sample digests the Na concentration with the ICP-OES at a wavelength of 589.592 nm. At this wavelength a (fitted) background correction is used.

Remarks:

3. A data management system and system controller are used. In this way the measurements are checked continuously and the data output is in concentration units in the digests.
4. Due to the linearity of the ICP-OES it is possible to calibrate the ICP-OES only with the highest and zero standard of the standard series. As a check all standards can be measured as samples.
5. Scandium (5 mg/L), at an wavelength of431.408 nm, is used in the author's laboratory as an internal standard to compensate for matrix effects.

## 9. CALCULATION

9.1 The total sodium content in the dried plant material, expressed in mmol/kg Na, is calculated by:

0.04350 * (a - b) * V / w

in which:

a is the concentration of sodium in the sample digest, in mg/L;

b is the concentration of sodium in the blank digest, in mg/L;

V is the total volume of digest at the end of the digestion procedure, in mL;

w is the weight of plant material sample, in g.

# 4.16 DETERMINATION OF NICKEL

# 4.16.A DETERMINATION OF NICKEL BY FLAME AAS

## 1. PRINCIPLE OF THE METHOD

1.1 The sample is nebulised into an air-acetylene flame, where it is vaporised; nickel compounds are atomised and the nickel atoms thus formed absorb radiation from a hollow cathode lamp. The absorbance is measured at a wavelength of 232.0 nm, using background correction.

1.2 This determination may be carried out on digest 2.3 (microwave digestion with $HNO_3$ - $H_2O_2$ - HF), digest 2.4 (digestion with $HNO_3$ - $H_2O_2$ - HF) and digest 2.7 (digestion by dry-ashing followed by treatment with HF).

## 2. RANGE AND DETECTION LIMIT

2.1 This procedure yields a standard curve that is linear up to approximately 2 mg/L Ni.

2.2 The detection limit is approximately 0.1 mg/L in the digest. The determination limit is approximately 0.3 mg/L (15-38 mg/kg in the dried plant material).

## 3. INTERFERENCES

3.1 To avoid interferences as much as possible, background correction is necessary.

## 4. PRECISION AND ACCURACY

4.1 The reproducibility of determinations by this procedure should give, at reasonable control and thorough sample preparation, a coefficient of variation within 5 % + 5 mg/kg.

## 5. APPARATUS

5.1 Atomic absorption spectrophotometer with a device for correcting background absorption.

## 6. REAGENTS

6.1A <u>Stock Solution, Ni concentration 1000 mg/L</u> - Merck nr 1.19792.

6.1B <u>Stock Solution, Ni concentration 1000 mg/L</u> - Dissolve 4.953 g nickel nitrate hexahydrate, $Ni(NO_3)_2.6H_2O$, in about 500 mL water in a 1000-mL volumetric flask and make up to volume.

6.2 Standard Solution, Ni concentration 10 mg/L - Pipette 1.00 mL stock solution (6.1A or 6.1B) into a 100-mL volumetric flask and make up to volume.

Remark:

1. $Ni(NO_3)_2.6H_2O$ may absorb water on standing. The reagent should be standardised by titration with EDTA at pH 10 with murexide as an indicator.

## 7. CALIBRATION AND STANDARDS

7.1 Standard Series - Pipette 0 – 1.00 – 2.00 – 5.00 – 10.00 – 15.00 – 20.00 mL of the standard solution (6.2) into 100-mL volumetric flasks, which already contain 40 mL water. Add either 10 mL concentrated nitric acid (65 %) (digestion 2.3), 3.0 mL concentrated nitric acid (65 %) (digestion 2.4) or 1.0 mL concentrated hydrochloric acid (36 %) (digestion 2.7). Let cool down and make up to the mark with water. This standard series has Ni concentrations of 0 – 0.1 – 0.2 – 0.5 – 1.0 – 1.5 – 2.0 mg/L.

7.2 Calibration Curve - The absorbance (A) is plot versus mg/L nickel in the standard series.

## 8. PROCEDURE

8.1 Measurement – Measure in the standard series, the blanks and sample digests the Ni concentration with flame AAS at a wavelength of 232.0 nm, using a just blue (stoichiometric) air-acetylene flame. Use scale expansion if necessary.

## 9. CALCULATION

9.1 The total nickel content of the dried plant material, expressed in μg/kg Ni, is calculated by:

$1000 * (a - b) * V / w$

in which:

a is the concentration of nickel in the sample digest, in mg/L;

b is the concentration of nickel in the blank digest, in mg/L;

V is the total volume of digest at the end of the digestion procedure, in mL;

w is the weight of plant material sample, in g.

# 4.16.B DETERMINATION OF NICKEL BY ICP-OES

## 1. PRINCIPLE OF THE METHOD

1.1 Solutions with nickel compounds are nebulised into an argon plasma, where all components are vaporised. Nickel compounds are dissociated and excited, and then emit radiation of which the intensity is measured at a wavelength of 231.604 nm.

1.2 This determination may be carried out on digest 2.3 (microwave digestion with $HNO_3$ - $H_2O_2$ - HF), digest 2.4 (digestion with $HNO_3$ - $H_2O_2$ - HF), digest 2.6 (digestion with $HNO_3$ - $HClO_4$ - $H_2SO_4$) and digest 2.7 (digestion by dry-ashing followed by treatment with HF).

## 2. RANGE AND DETECTION LIMIT

2.1 This procedure yields a standard curve that is linear up to at least 2 mg/L Ni.

2.2 The detection limit is approximately 0.006 mg/L in the digest. The determination limit is approximately 0.018 mg/L (2 respectively 6 mg/kg in the dried plant material).

## 3. INTERFERENCES

3.1 No interferences are expected.

## 4. PRECISION AND ACCURACY

4.1 The reproducibility of determinations by this procedure should give, at reasonable control and thorough sample preparation, a coefficient of variation within 5 %.

## 5. APPARATUS

5.1 Inductively coupled plasma optical emission spectrometer.

## 6. REAGENTS

6.1A Stock Solution, Ni concentration 1000 mg/L - Merck nr 1.02640.

6.1B Stock Solution, Ni concentration 1000 mg/L - Dissolve 4.953 g nickel nitrate hexahydrate, $Ni(NO_3)_2.6H_2O$, in about 500 mL ultra pure water in a 1000-mL volumetric flask and make up to volume.

6.2 Standard Solution, Ni concentration 100 mg/L - Pipette 1 mL concentrated nitric acid (65 % s.p.) in a 100-mL polythene volumetric flask which already contains about 50 mL ultra pure water. Add 10.00 mL stock solution (6.1A or 6.1B) and make up to volume with ultra pure water.

# 7. CALIBRATION AND STANDARDS

7.1 Standard Series - Pipette 0 – 1.00 – 2.00 mL of the stock solution (6.1A or 6.1B) into 100-mL volumetric flasks, which already contain 40 mL water. Add either 10 mL concentrated nitric acid (65 %) (digestion 2.3), 3.0 mL concentrated nitric acid (65 %) (digestion 2.4) or 1.0 mL concentrated hydrochloric acid (36 %) (digestion 2.7). Let cool down and make up to the mark with water. This standard series has Ni concentrations of 0 – 1 – 2 mg/L.

7.2 Calibration Curve - The emission counts are plot versus mg/L nickel in the standard series.

Remark:

1. A mix standard series can be used for simultaneous measurements with ICP-OES (Al, Ca, Cd, Co, Cr, Cu, Fe, K, Mg, Mn, Na, Ni, P, S and Zn).

# 8. PROCEDURE

8.1 Measurement - Measure in the standard series, the blanks and the sample digests the Ni concentration with the ICP-OES at a wavelength of 231.604 nm. At this wavelength a (left and right side) background correction is used.

Remarks:

2. A data management system and system controller are used. In this way the measurements are checked continuously and the data output is in concentration units in the digests.
3. Due to the linearity of the ICP-OES it is possible to calibrate the ICP-OES only with the highest and zero standard of the standard series. As a check all standards can be measured as samples.
4. Scandium (5 mg/L), at an wavelength of 255.235 nm, is used in the author's laboratory as an internal standard to compensate for matrix effects.

# 9. CALCULATION

9.1 The total nickel content in the dried plant material, expressed in mg/kg Ni, is calculated by:

$(a - b) * V / w$

in which:

a is the concentration of nickel in the sample digest, in mg/L;

b is the concentration of nickel in the blank digest, in mg/L;

V is the total volume of digest at the end of the digestion procedure, in mL;

w is the weight of plant material sample, in g.

# 4.16.C DETERMINATION OF NICKEL BY ICP-MS

## 1. PRINCIPLE OF THE METHOD

1.1 Solutions with nickel compounds are nebulised into an argon plasma, where all components are vaporised. The ions produced are entrained in the plasma gas and introduced into a mass spectrometer, sorted according to their mass-to-charge ratios, and quantified with a channel electron multiplier. Nickel is determined at mass 60 or 62 amu.

1.2 This determination may be carried out on digest 2.3 (microwave digestion with $HNO_3$ - $H_2O_2$ - HF), digest 2.4 (digestion with $HNO_3$ - $H_2O_2$ - HF) and digest 2.7 (digestion by dry-ashing followed by treatment with HF).

## 2. RANGE AND DETECTION LIMIT

2.1 This procedure yields a standard curve that is linear up to at least 50 μg/L Ni.

2.2 The detection limit is approximately 0.02 μg/L in the digest. The determination limit is approximately 0.06 μg/L (7 respectively 21 μg/kg in the dried plant material).

## 3. INTERFERENCES

3.1 The $^{44}Ca^{16}O$ molecule has the same mass as $^{60}Ni$. A correction factor can be applied. The $^{46}Ti^{16}O$ molecule has the same mass as $^{62}Ni$. A correction factor can be applied.

## 4. PRECISION AND ACCURACY

4.1 The reproducibility of determinations by this procedure should give, at reasonable control and thorough sample preparation, a coefficient of variation within 10 %.

## 5. APPARATUS

5.1 Inductively coupled plasma mass spectrometer.

## 6. REAGENTS

6.1A Stock Solution, Ni concentration 1000 mg/L - Merck nr 1.02640.

6.1B Stock Solution, Ni concentration 1000 mg/L - Dissolve 4.953 g nickel nitrate hexahydrate, $Ni(NO_3)_2.6H_2O$, in about 500 mL ultra pure water in a 1000-mL volumetric flask and make up to volume.

6.2 Standard Solution, Ni concentration 1 mg/L - Pipette 1 mL concentrated nitric acid (65 % s.p.) in a 1000-mL polythene volumetric flask which already contains about 500 mL ultra pure water. Add 1.00 mL stock solution (6.1A or 6.1B) and make up to volume with ultra pure water.

Remark:

1. $Ni(NO_3)_2.6H_2O$ may absorb water on standing. The reagent should be standardised by titration with EDTA at pH=10 with Murexide as an indicator.

# 7. CALIBRATION AND STANDARDS

7.1 Standard Series - Pipette 0 – 1.00 – 2.00 – 5.00 mL of the standard solution (6.2) in about 40 mL ultra pure water in a 100-mL polythene volumetric flasks. Add either 1.0 mL concentrated nitric acid (65 %) (digestion 2.3), 0.3 mL concentrated nitric acid (65 %) (digestion 2.4) or 0.1 mL concentrated hydrochloric acid (36 %) (digestion 2.7). Let cool down and make up to the mark with ultra pure water. This standard series has Ni concentrations of 0 –10 – 20 – 50 µg/L.

7.2 Calibration Curve - The counts per second are plot versus µg/L nickel in the standard series.

Remark:

2. A mix standard series can be used for simultaneous measurements with ICP-MS (Al, As, Cd, Co, Cr, Cu, Mn, Ni, Pb, Sb, Sn, V and Zn).

# 8. PROCEDURE

8.1 Measurement – Measure in the standard series, the blank digests and sample digests the Ni concentration with the ICP-MS at a mass of 60 or 62 amu. Use a correction factor if necessary.

Remarks:

3. A data management system and system controller are used. In this way the measurements are checked continuously and the data output is in concentration units in the digests.
4. Due to the linearity of the ICP-MS it is possible to calibrate the ICP-MS only with the highest and zero standard of the standard series. As a check all standards can be measured as samples.

# 9. CALCULATION

9.1 The total nickel content in the dried plants material, expressed in µg/kg Ni, is calculated by:

$(a - b) * V / w$

in which:

a is the concentration of nickel in the sample digest, in µg/L;

b is the concentration of nickel in the blank digest, in µg/L;

V is the total volume of digest at the end of the digestion procedure, in mL;

w is the weight of plant material sample, in g.

# 4.17 DETERMINATION OF PHOSPHORUS

## 4.17.A DETERMINATION OF TOTAL PHOSPHORUS BY SPECTROPHOTOMETRY

## 1. PRINCIPLE OF THE METHOD

1.1 In an acidic medium, orthophosphates form a yellow-coloured complex with molybdate ions. The combination of P, Sb and Mo leads after reduction with ascorbic acid to the formation of a very stable blue-coloured phosphormolybdenum complex. At a wavelength of 880 nm a very sensitive and stable measurement of P is possible.

1.2 This determination may be carried out on digest 2.1 (digestion with $H_2SO_4$ - salicylic acid - $H_2O_2$ - Se) and digest 2.2 (digestion with $H_2SO_4$ - salicylic acid - $H_2O_2$).

## 2. RANGE AND DETECTION LIMIT

2.1 This procedure yields a standard curve that is linear up to approximately 5 mg/L $PO_4$.

2.2 The detection limit is approximately 0.03 mg/L in the digest. The determination limit is approximately 0.1 mg/L (0.2 respectively 0.5 mmol/kg in the dried plant material).

## 3. INTERFERENCES

3.1 Arsenic and Silica - $As^{5+}$ and $Si^{4+}$ form similarly coloured complexes under these conditions. Both interferences are negligible at the low concentrations, which are normally present in these plant material digests.

Remark:

1. The blue colour varies with the redox conditions of the medium and with pH.

## 4. PRECISION AND ACCURACY

4.1 The reproducibility of determinations by this procedure should give, at reasonable control and thorough sample preparation, a coefficient of variation within 5 % + 5 mmol/kg.

## 5. APPARATUS

5.1 Spectrophotometer equipped with a flow through cell.

## 6. REAGENTS

6.1A Stock Solution, $PO_4$ concentration 1000 mg/L - Merck nr 1.19898.

6.1B Stock Solution, $PO_4$ concentration 1000 mg/L - Dissolve 1.432 g potassium dihydrogen phosphate, $KH_2PO_4$ (see remark 2), in about 900 mL water in a volumetric flask of 1000 mL. Make up to 1000 mL with water.

6.2 Ascorbic Acid Solution - Dissolve 1.76 g ascorbic acid, $C_6H_8O_6$, in 100 mL ultra pure water and mix. Prepare fresh daily.

6.3 Ammonium Molybdate Solution - Dissolve 40 g ammonium molybdate tetrahydrate, $(NH_4)_6Mo_7O_{24}.4H_2O$, in ultra pure water and make up to 1000 mL. This solution should be stored in a bottle made of hard glass.

6.4 Potassium Antimonyl Tartrate Solution - Dissolve 0.274 g potassium antimonyl tartrate, $KSbOC_4H_4O_6.5H_2O$, in ultra pure water and make up to 100 mL with ultra pure water.

6.5 Sulphuric Acid Solution 2.5 mol/L - Dilute carefully, in portions, 140 mL concentrated sulphuric acid (96 %) in about 500 mL ultra pure water in a 1000-mL volumetric flask. Allow the mixture to cool off and make up to volume with ultra pure water.

6.6 Anti-coagulation Agent - Wetting agent Aerosol 22, Merck nr 13908.

6.7 Mixed Reagent - Add successively with a graduated cylinder and mix after each addition: 50 mL sulphuric acid (6.5), 15 mL ammonium molybdate solution (6.3), 30 mL ascorbic acid solution (6.2) and 5 mL potassium antimonyl tartrate solution (6.4). Prepare fresh daily.

6.8 Diluted Mixed Reagent - Mix 80 mL of the mixed reagent (6.6) with 300 mL ultra pure water; if samples of digestion 2.1 (digestion with $H_2SO_4$ - salicylic acid - $H_2O_2$ - Se) are to be measured, add 0.5 mL of anti-coagulating agent (6.6).

Remarks:

2. The potassium dihydrogen phosphate has to be dried at 105 °C for at least 2 hours just before weighing. Just before drying, any lumps should be cut down so that only fine crystals remain. This is important, because coarse crystals contain much occluded water in their cavities and thus would need a higher drying temperature. The fine crystals should, however, not be pulverised because this enhances deliquescence (= attraction of water) and/or efflorescence (= weathering) later on. After the recommended heating procedure, the potassium dihydrogen phosphate should be cooled in a desiccator containing preferably magnesium perchlorate as a desiccant. The dried potassium dihydrogen phosphate should be weighed as soon as it has reached the ambient temperature.
3. For this determination, all solutions must be prepared with ultra pure water, since demineralised water may contain varying amounts of Si.
4. This solution should be stored in a bottle made of hard glass.
5. For digest 2.2 (digestion with $H_2SO_4$ - salicylic acid - $H_2O_2$), where no Se is used, the anti-coagulating agent is better left out because it slows down the colour formation

# 7. CALIBRATION AND STANDARDS

7.1 Standard Series - Pipette 0 – 1.00 – 2.00 – 3.00 – 4.00 – 5.00 mL of the stock solution (6.1A or 6.1B) into 100-mL volumetric flasks, which already contain 40 mL ultra pure water. Add 4.5 mL concentrated sulphuric acid (96 %) (digestion 2.1 or 2.2). Let cool down and make up to the mark with ultra pure water. This standard series has $PO_4$ concentrations of 0 – 10 – 20 – 30 – 40 – 50 mg/L.

7.2 Calibration Curve - The absorbance (A) is plot versus mg/L phosphate in the standard series.

# 8. PROCEDURE

8.1 Measurement - Dilute the standard series, the blanks and all digests 1+9 (v/v) with ultra pure water. Pipette 1.0 mL of the diluted standard series, the diluted blanks and the diluted sample digests into test tubes. Add 3.8 mL of the diluted mixed reagent (6.8) and mix. Allow to stand for 10 min (digest 2.2) or 1 hour (digest 2.1). Measure the absorbance in a 1-cm cuvette at a wavelength of 880 nm.

Remarks:

6. The phosphomolybdenum blue complex reaches its maximum intensity after 1 hour and is stable for at least 10 hours.
7. Only part of the $Mo^{6+}$ will be reduced. Hence, samples that have high phosphate concentrations cannot be diluted once that the blue colour has been formed.

# 9. CALCULATION

9.1 The total phosphorus content in the dried plant material, expressed in mmol/kg P, is calculated by:

0.01053 * (a - b) * V / w

in which:

a is the concentration of phosphorus in the diluted sample digest, in mg/L;
b is the concentration of phosphorus in the diluted blank digest, in mg/L;
V is the total volume of digest at the end of the digestion procedure, in mL;
w is the weight of plant material sample, in g.

# 4.17.B DETERMINATION OF TOTAL PHOSPHORUS BY SFA

## 1. PRINCIPLE OF THE METHOD

1.1 Orthophosphates form a yellow-coloured complex with molybdate ions in an acid medium. After addition of ascorbic acid and Sb, a blue-coloured phosphomolybdenum complex is formed, of which the absorbance is measured at a wavelength of 660 nm. The determination is carried out as a so-called segmented-flow analysis (SFA).

1.2 This determination may be carried out on digest 2.1 (digestion with $H_2SO_4$ - salicylic acid - $H_2O_2$ - Se) and digest 2.2 (digestion with $H_2SO_4$ - salicylic acid - $H_2O_2$).

Remark:

1. Sb is added to complete the reduction of the yellow phosphomolybdenum complex to the blue phosphomolybdenum complex.

## 2. RANGE AND DETECTION LIMIT

2.1 This procedure yields a standard curve that is linear up to approximately 60 mg/L P.

2.2 The detection limit is approximately 0.05 mg/L in the digest. The determination limit is approximately 0.15 mg/L (0.8 mmol/kg in the dried plant material).

## 3. INTERFERENCES

3.1 $As^{5+}$ and $Si^{4+}$ form similarly coloured complexes under these conditions. Both interferences are negligible at the low concentrations, which are normally present in these plant material digests.

## 4. PRECISION AND ACCURACY

4.1 The reproducibility of determinations by this procedure should give, at reasonable control and thorough sample preparation, a coefficient of variation within 5 % + 5 mmol/kg.

## 5. APPARATUS

5.1 Segmented-flow analysis system (sampler, pump, $PO_4$-unit, photometer, computer). In the authors' laboratory, a Skalar Segmented Flow Analyzer is used, but any SFA system will do. Note that with other systems the flow diagram may require adaptation.

# 6. REAGENTS

6.1 Stock Solution, P concentration 1000 mg/L - Dissolve 4.394 g potassium dihydrogen o-phosphate, $KH_2PO_4$ (see remark 2) in a 1000-mL volumetric flask which already contains about 800 ultra pure water and make up to volume.

6.2 Potassium Antimony Tartrate Solution - Dissolve 0.300 g potassium antimony tartrate, $K(SbO)C_4H_4H_6.0.5H_2O$, in a 100-mL volumetric flask which already contains about 80 mL ultra pure water and make up to volume. This solution is stable for one month at 4 °C.

6.7 Ascorbic Acid - Dissolve 18 g ascorbic acid, $C_6H_8O_6$, in a 1000-mL volumetric flask which already contains about 800 mL ultra pure water. Add 20 mL stock solution potassium antimony tartrate (6.2) and make up to volume. This solution is stable for one week at 4 °C.

6.8 Sulphuric Acid Solution - Pipette 40 mL concentrated sulphuric acid (96 %) in a 1000-mL volumetric flask, which already contains about 800 mL ultra pure water. Let cool down and make up to volume then add 2 mL FFD6, sodium dodecyl diphenyloxide disulphonate (45-47%),and mix.

6.5 Diluted FFD6 - Pipette 2 mL FFD6, sodium dodecyl diphenyloxide disulphonate (45-47%), in a 1000-mL volumetric flask which already contains about 800 mL ultra pure water and make up to volume.

6.6 Ammonium Molybdate - Pipette 40 mL concentrated sulphuric acid (96 %) in a 1000-mL volumetric flask which contains already about 800 mL ultra pure water and let cool down. Hereafter add 4.8 g ammonium molybdate, $(NH_4)_6Mo_7O_{24}.4H_2O$, and dissolve. Make up to volume, add 2 mL FFD6, sodium dodecyl diphenyloxide disulphonate (45-47%), and mix well (see remark 3).

6.7 Rinsing Liquid Sampler - Pipette 33 mL concentrated sulphuric acid (96 %) in a 1000-mL volumetric flask, which contains already about 800 mL ultra pure water. Allow cooling and make up to volume.

Remark:

2. The potassium dihydrogen phosphate has to be dried at 105 °C for at least 2 hours just before weighing. Just before drying, any lumps should be cut down so that only fine crystals remain. This is important, because coarse crystals contain much occluded water in their cavities and thus would need a higher drying temperature. The fine crystals should, however, not be pulverised because this enhances deliquescence (= attraction of water) and/or efflorescence (= weathering) later on. After the recommended heating procedure, the potassium dihydrogen phosphate should be cooled in a desiccator containing preferably magnesium perchlorate as a desiccant. The dried potassium dihydrogen phosphate should be weighed as soon as it has reached the ambient temperature.
3. The ammonium molybdate chemicals must not be in contact with metal components.

# 7. CALIBRATION AND STANDARDS

7.1 Standard Series - Pipette into 100-mL volumetric flasks, that contain already about 50 mL water, 4.5 mL of concentrated sulphuric acid (96 %). Allow to cool. Then add 0 – 2.00 – 4.00 – 6.00 mL of the stock solution (6.1) and dilute to volume. The standard series has P concentrations of 0 – 20 – 40 – 60 mg/L.

7.2 Calibration Curve - The absorbance (A) is plot versus mg/L phosphorus in the standard series.

# 8. PROCEDURE

8.1 Measurement - Start the segmented-flow system according to the scheme given in figure 1. Measure the absorbance of the standard series, the blanks and the sample digests at a wavelength of 660 nm.

Remarks:

4. The absorbance is measured at 660 nm because the system is provided with a filter of that wavelength. Measurements may, however, be taken at 720 nm or 880 nm as well.
5. A data management system and system controller are used. In this way the measurements are checked continuously and the data output is in concentration units in the digests.

# 9. CALCULATION

9.1 The total phosphorus content in the dried plant material, expressed in mmol/kg P, is calculated by:

0.03229 * (a - b) * V / w

in which:

a is the concentration of phosphorus in the sample digest, in mg/L;
b is the concentration of phosphorus in the blank digest, in mg/L;
V is the total volume of digest at the end of the digestion procedure, in mL;
w is the weight of plant material sample, in g.

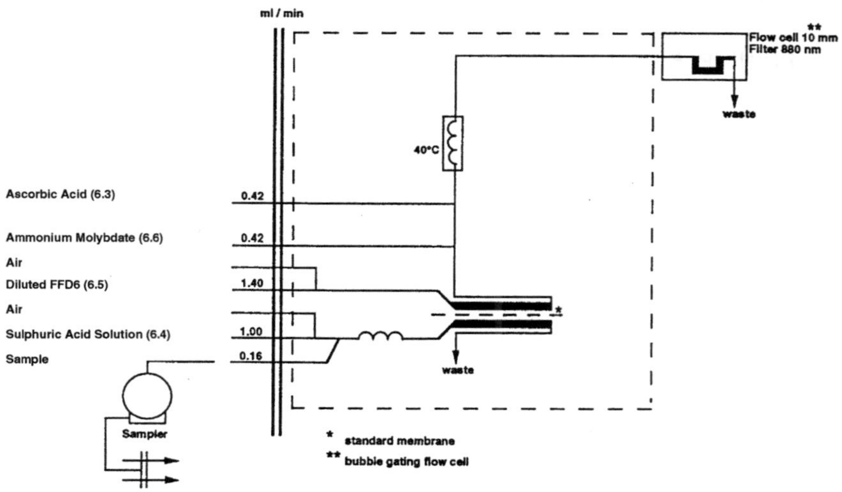


Figure 1 Flow diagram for the determination of phosphorus by SFA.

# 4.17.C DETERMINATION OF TOTAL PHOSPHORUS BY ICP-OES

## 1. PRINCIPLE OF THE METHOD

1.1 Solutions with phosphorus compounds are nebulised into an argon plasma, where all components are vaporised. Phosphorus compounds are dissociated and excited, and then emit radiation of which the intensity is measured at a wavelength of 178.222 nm.

1.2 This determination may be carried out on digest 2.1 (digestion with $H_2SO_4$ - salicylic acid - $H_2O_2$ - Se) and digest 2.2 (digestion with $H_2SO_4$ - salicylic acid - $H_2O_2$), digest 2.3 (microwave digestion with $HNO_3$ - $H_2O_2$ - HF), digest 2.4 (digestion with $HNO_3$ - $H_2O_2$ - HF) and digest 2.7 (digestion by dry-ashing followed by treatment with HF).

## 2. RANGE AND DETECTION LIMIT

2.1 This procedure yields a standard curve that is linear up to at least 100 mg/L P.

2.2 The detection limit is approximately 0.04 mg/L in the digest. The determination limit is approximately 0.12 mg/L (0.4 respectively 1.2 mmol/kg in the dried plant material).

## 3. INTERFERENCES

3.1 No interferences are expected.

## 4. PRECISION AND ACCURACY

4.1 The reproducibility of determinations by this procedure should give, at reasonable control and thorough sample preparation, a coefficient of variation within 5 % + 5 mmol/kg.

## 5. APPARATUS

5.1 Inductively coupled plasma optical emission spectrometer.

## 6. REAGENTS

6.1A Stock Solution, P concentration 2000 mg/L – LPS Benelux nr 11289.

6.1B Stock Solution, $PO_4$ concentration 2000 mg/L - Dissolve 2.864 g potassium dihydrogen phosphate, $KH_2PO_4$ (see remark 1), in about 900 mL water in a volumetric flask of 1000 mL. Make up to 1000 mL with water.

Remark:

1. The potassium dihydrogen phosphate has to be dried at 105 ºC for at least 2 hours just before weighing. Just before drying, any lumps should be cut down so that only fine crystals remain. This is important, because coarse crystals contain much occluded water in their cavities and thus would need a higher drying temperature. The fine crystals should, however, not be pulverised because this enhances deliquescence (= attraction of water) and/or efflorescence (= weathering) later on. After the recommended heating procedure, the potassium dihydrogen phosphate should be cooled in a desiccator containing preferably magnesium perchlorate as a desiccant. The dried potassium dihydrogen phosphate should be weighed as soon as it has reached the ambient temperature.

# 7. CALIBRATION AND STANDARDS

7.1 Standard Series - Pipette 0 – 1.00 – 2.00 – 10.00 mL of the stock solution (6.1A or 6.1B) into 100-mL volumetric flasks which already contain 40 mL water. Add either 10 mL concentrated nitric acid (65 %) (digestion 2.3), 3.0 mL concentrated nitric acid (65 %) (digestion 2.4) or 1.0 mL concentrated hydrochloric acid (36 %) (digestion 2.7). Let cool down and make up to the mark with water. This standard series has $PO_4$ concentrations of 0 – 10 – 20 – 100 mg/L.

7.2 Calibration Curve - The emission counts are plot versus mg/L phosphorus in the standard series.

Remark:

2. A mix standard series can be used for simultaneous measurements with ICP-OES (Al, Ca, Cd, Co, Cr, Cu, Fe, K, Mg, Mn, Na, Ni, P, S and Zn).

# 8. PROCEDURE

8.1 Measurement - Measure in the standard series, the blanks and the sample digests the P concentration with the ICP-OES at a wavelength of 178.222 nm. At this wavelength a (fitted) background correction is used.

Remarks:

3. A data management system and system controller are used. In this way the measurements are checked continuously and the data output is in concentration units in the digests.
4. Due to the linearity of the ICP-OES it is possible to calibrate the ICP-OES only with the highest and zero standard of the standard series. As a check all standards can be measured as samples.
5. Scandium (5 mg/L), at a wavelength of 255.235 nm, is used in the author's laboratory as an internal standard to compensate for matrix effects.

# 9. CALCULATION

9.1 The total phosphorus content in the dried plant material, expressed in mmol/kg P, is calculated by:

0.03229 * (a - b) * V / w

in which:

a is the concentration of phosphorus in the sample digest, in mg/L;

b is the concentration of phosphorus in the blank digest, in mg/L;
V is the total volume of digest at the end of the digestion procedure, in mL;
w is the weight of plant material sample, in g.

# 4.18 DETERMINATION OF LEAD

# 4.18.A DETERMINATION OF LEAD BY FLAME AAS

## 1. PRINCIPLE OF THE METHOD

1.1 The sample is nebulised into an air-acetylene flame, where it is vaporised; lead compounds are atomised and the lead atoms thus formed absorb radiation from a hollow cathode lamp. The absorbance is measured at a wavelength of 217.0 nm, using deuterium background correction.

1.2 This determination may be carried out on digest 2.3 (microwave digestion with $HNO_3$ - $H_2O_2$ - HF), digest 2.4 (digestion with $HNO_3$ - $H_2O_2$ - HF), digest 2.6 (digestion with $HNO_3$ - $HClO_4$ - $H_2SO_4$) and digest 2.7 (digestion by dry-ashing followed by treatment with HF).

## 2. RANGE AND DETECTION LIMIT

2.1 This procedure yields a standard curve that is linear up to approximately 2 mg/L Pb.

2.2 The detection limit is approximately 0.05 mg/L in the digest. The determination limit is approximately 0.15 mg/L (7.5-19 mg/kg in the dried plant material).

## 3. INTERFERENCES

3.1 To avoid interferences as much as possible, background correction is necessary.

## 4. PRECISION AND ACCURACY

4.1 The reproducibility of determinations by this procedure should give, at reasonable control and thorough sample preparation, a coefficient of variation within 10 % + 10 mg/kg.

## 5. APPARATUS

5.1 Flame atomic absorption spectrophotometer with a device for correcting background absorption.

## 6. REAGENTS

6.1A Stock Solution, Pb concentration 1000 mg/L - Merck nr 1.19776.

6.1B Stock Solution, Pb concentration 1000 mg/L - Dissolve 1.5985 g lead nitrate, $Pb(NO_3)_2$, in some water in a 1000-mL volumetric flask and make up to volume.

6.2 Standard Solution, Pb concentration 100 mg/L - Pipette 10.00 mL stock solution (6.1A or 6.1B) into a 100-mL volumetric flask and make up to volume.

Remark:

1. $Pb(NO_3)_2$ may absorb water on standing. The reagent should be standardised by titration with EDTA at pH 10 in the presence of tartrate ions with Eriochrome Black T as an indicator.

## 7. CALIBRATION AND STANDARDS

7.1 Standard Series - Pipette 0 – 0.50 – 1.00 – 2.00 – 3.00 – 4.00 mL of the standard solution (6.2) into 100-mL volumetric flasks, which already contain 40 mL, water. Add either 10 mL concentrated nitric acid (65 %) (digestion 2.3), 3.0 mL concentrated nitric acid (65 %) (digestion 2.4), 0.45 concentrated sulphuric acid (96 %) (digestion 2.6) or 1.0 mL concentrated hydrochloric acid (36 %) (digestion 2.7). Let cool down and make up to the mark with water. This standard series has Pb concentrations of 0 – 0.5 – 1.0 – 2.0 – 3.0 – 4.0 mg/L.

7.2 Calibration Curve - The absorbance (A) is plot versus mg/L lead in the standard series.

## 8. PROCEDURE

8.1 Measurement - Measure in the standard series, the blanks and the sample digests the Pb concentration with flame AAS, at a wavelength of 217.0 nm. Use scale expansion if necessary.

Remark:

2. Samples with high iron content may give too low results when applying a deuterium background correction system (Van der Lee et al., 1987). In this case Smith-Hieftje or Zeeman backgroun correction should be used.

## 9. CALCULATION

9.1 The total lead content in the dried plant material, expressed in μg/kg Pb, is calculated by:

$1000 * (a - b) * V / w$

in which:

a is the concentration of lead in the sample digest, in mg/L;
b is the concentration of lead in the blank digest, in mg/L;
V is the total volume of digest at the end of the digestion procedure, in mL;
w is the weight of plant material sample, in g.

## 10. REFERENCES

10.1 Van der Lee, J.J., E.J.M. Temminghoff, V.J.G. Houba and I. Novozamsky. 1987. Background corrections in the determination of Cd and Pb by flame AAS in plant and soil samples with high Fe levels. Appl. Spectrosc. 41: 388-390.

# 4.18.B DETERMINATION OF LEAD BY ICP-OES

## 1. PRINCIPLE OF THE METHOD

1.1 Solutions with lead compounds are nebulised into an argon plasma, where all components are vaporised. Lead compounds are dissociated and excited, and then emit radiation of which the intensity is measured at a wavelength of 220.353 nm.

1.2 This determination may be carried out on digest 2.3 (microwave digestion with $HNO_3$ - $H_2O_2$ - HF), digest 2.4 (digestion with $HNO_3$ - $H_2O_2$ - HF) and digest 2.7 (digestion by dry-ashing followed by treatment with HF).

## 2. RANGE AND DETECTION LIMIT

2.1 This procedure yields a standard curve that is linear up to at least 2 mg/L Pb.

2.2 The detection limit is approximately 0.04 mg/L in the digest. The determination limit is approximately 0.12 mg/L (13 respectively 40 mg/kg in the dried plant material).

## 3. INTERFERENCES

3.1 No interferences are expected.

## 4. PRECISION AND ACCURACY

4.1 The reproducibility of determinations by this procedure should give, at reasonable control and thorough sample preparation, a coefficient of variation within 10 % + 10 mg/kg.

## 5. APPARATUS

5.1 Inductively coupled plasma optical emission spectrometer.

## 6. REAGENTS

6.1A <u>Stock Solution, Pb concentration 1000 mg/L</u> - Merck nr 1.19776.

6.1B <u>Stock Solution, Pb concentration 1000 mg/L</u> - Dissolve 1.5985 g lead nitrate, $Pb(NO_3)_2$, in some water in a 1000-mL volumetric flask and make up to volume.

6.2 <u>Standard Solution, Pb concentration 40 mg/L</u> - Pipette 4.00 mL stock solution (6.1A or 6.1B) in a 100-mL volumetric flask and make up to volume with water.

<u>Remark</u>:

1. $Pb(NO_3)_2$ may absorb water on standing. The reagent should be standardised by titration with EDTA at pH 10 in the presence of tartrate ions with Eriochrome Black T as an indicator.

## 7. CALIBRATION AND STANDARDS

7.1 Standard Series - Pipette 0 – 1.00 – 2.00 mL of the standard solution (6.2) into 100-mL volumetric flasks which already contain 40 mL water. Add either 10 mL concentrated nitric acid (65 %) (digestion 2.3), 3.0 mL concentrated nitric acid (65 %) (digestion 2.4) or 1.0 mL concentrated hydrochloric acid (36 %) (digestion 2.7). Let cool down and make up to the mark with water. This standard series has Pb concentrations of 0 – 1.0 – 2.0 mg/L.

7.2 Calibration Curve - The emission counts are plot versus mg/L lead in the standard series.

Remark:

2. For simultaneous measurements with ICP-OES a mix standard series has to be used without Pb (Al, Ca, Cd, Cu, Fe, K, Mg, Mn, Na, P, S and Zn), since Pb and S ($SO_4$) can not be used in one mix standard due to $PbSO_4$ precipitation.

## 8. PROCEDURE

8.1 Measurement - Measure in the standard series, the blanks and the sample digests the Pb concentration with the ICP-OES at a wavelength of 220.353 nm. At this wavelength a (left and right side) background correction is used.

Remarks:

3. A data management system and system controller are used. In this way the measurements are checked continuously and the data output is in concentration units in the extracts.
4. Due to the linearity of the ICP-OES it is possible to calibrate the ICP-OES only with the highest and zero standard of the standard series. As a check all standards can be measured as samples.
5. Scandium (5 mg/L), at a wavelength of 255.235 nm, is used in the author's laboratory as an internal standard to compensate for matrix effects.

## 9. CALCULATION

9.1 The total lead content in the dried plant material, expressed in mg/kg Pb, is calculated by:

$(a - b) * V / w$

in which:

a is the concentration of lead in the sample digest, in mg/L;

b is the concentration of lead in the blank digest, in mg/L;

V is the total volume of digest at the end of the digestion procedure, in mL;

w is the weight of plant material sample, in g.

# 4.18.C DETERMINATION OF LEAD BY ETA-AAS

## 1. PRINCIPLE OF THE METHOD

1.1 Lead ions in the digest are subsequently dried, ashed and vaporised by electrical heating in a graphite furnace. The lead atoms thus formed absorb radiation from a hollow cathode lamp. The absorbance is measured at a wavelength of 283.3 nm.

1.2 This determination may be carried out on digest 2.3 (microwave digestion with $HNO_3$ - $H_2O_2$ - HF), digest 2.4 (digestion with $HNO_3$ - $H_2O_2$ - HF) and digest 2.7 (digestion by dry-ashing followed by treatment with HF).

## 2. RANGE AND DETECTION LIMIT

2.1 This procedure yields a standard curve that is linear up to approximately 50 µg/L Pb.

2.2 The detection limit is approximately 2.8 µg/L in the digest. The determination limit is approximately 8.4 µg/L (400-1000 µg/kg in the dried plant material).

## 3. INTERFERENCES

3.1 To avoid interferences as much as possible, background correction is necessary.

Remark:

1. Alternatively the 217.0 nm line (more sensitive) can be used. However, non-specific absorption is than more important and background correction has to be used.

## 4. PRECISION AND ACCURACY

4.1 The reproducibility of determinations by this procedure should give, at reasonable control and thorough sample preparation, a coefficient of variation within 10 %.

## 5. APPARATUS

5.1 Electrothermal atomisation (graphite furnace) atomic absorption spectrometer with a device for correcting background absorption.

5.2 Polythene cups.

## 6. REAGENTS

6.1A Stock Solution, Pb concentration 1000 mg/L - Merck nr 1.19776.

6.1B Stock Solution, Pb concentration 1000 mg/L - Dissolve 1.5985 g lead nitrate, $Pb(NO_3)_2$, in some ultra pure water in a 1000-mL volumetric flask and make up to volume.

6.2 Diluted Standard Solution, Pb concentration 1 mg/L - Pipette 1 mL concentrated nitric acid (65 % s.p.) in a 1000-mL polythene volumetric flask which already contains about 500 mL ultra pure water. Add 1.00 mL stock solution (6.1A or 6.1B) and make up to volume with ultra pure water.

6.3 Nitric Acid Solution 5 mol/L - Dilute 34.7 mL concentrated nitric acid (65 %) in about 30 mL ultra pure water in a 100-mL volumetric flask. Let cool down and dilute to volume.

6.4 Matrix modifier, Palladium(II)chloride 0.2 % - Dissolve 0.20 g palladium(II) chloride in 0.5 mL concentrated nitric acid (65 %) and heat to dissolve. Heat till almost dry and transfer the solution into a 100-mL volumetric flask and make up to volume.

6.5 Butanol.

6.6 Acidified Triton-X 100 Solution 1 % - Dissolve 1.00 g triton-X in about 20 mL ultra pure water. Transfer the solution into a 100-mL volumetric flask, add 20 mL nitric acid solution (6.3) and make up to volume.

6.7 Propanol-2 Solution 5 % - Dilute 25 mL propanol-2 in some ultra pure water in a 500-mL volumetric flask and make up to volume.

Remark:

2. $Pb(NO_3)_2$ may absorb water on standing. The reagent should be standardised by titration with EDTA at pH 10 in the presence of tartrate ions with Eriochrome Black T as an indicator.

## 7. CALIBRATION AND STANDARDS

7.1 Standard Series - Pipette 0 – 1.00 – 2.00 – 3.00 – 4.00 – 5.00 mL of the diluted standard solution (6.2) into 100-mL volumetric flasks which already contain 40 mL ultra pure water. Add either 10 mL concentrated nitric acid (65 %) (digestion 2.3), 3.0 mL concentrated nitric acid (65 %) (digestion 2.4) or 1.0 mL concentrated hydrochloric acid (36 %) (digestion 2.7). Let cool down and make up to the mark with ultra pure water. This standard series has Pb concentrations of 0 – 10 – 20 – 30 – 40 – 50 μg/L.

7.2 Calibration Curve - The absorbance (A) is plot versus mg/L lead in the standard series.

## 8. PROCEDURE

8.1 Measurement - Pipette 1.00 mL of the standard series, blanks and sample digests into polythene cups that fit in the automatic sampler of the atomic absorption spectrophotometer. Add 0.05 mL of the acidified triton-X solution (6.6) and mix thoroughly with an electric mini-stirrer. Pipette 1 mL of the matrix modifier (6.4) into another cup, add 0.05 mL of butanol (6.5) and mix. Put standards, modifier, blanks and sample digests in the appropriate place of the sampler. Heat in a graphite furnace according to an appropriate time-temperature programme (see remarks 6 and 7). Measure the absorbance at 283.3 nm in the atomisation phase (use background correction).

Remarks:

3. Samples with high iron content may give too low results when applying a deuterium background correction system (Van der Lee et al., 1987).
4. Every sample should be measured at least three times and for calculation the mean can be used. The relative standard deviation should be less than 2 % for three replicates.
5. The wash solution of the automatic sampler contains a 5 % propanol-2 solution (6.7) in order to lower its surface tension and to prevent growth of bacteria.
6. The measurements can be performed with any ETA-AAS system. The present method was worked out using a Varian SpectrAA-300 atomic absorption spectrometer equipped with a graphite tube atomiser, an automatic sampler and a Zeeman-effect background correction system and for a matrix of 0.4 M $HNO_3$ (see remark 7). The operating parameters and temperature programme are given below. Pyrolytically coated partition tubes are used in the author's laboratory. For using the given temperature program 0.2 % palladium chloride in 5 % butanol as matrix modifier is necessary. Butanol (0.05 mL) is added, to 1 mL of the Pd solution, before use, in order to achieve more reproducible drying conditions in the graphite atomiser (Temminghoff, 1990). The sample volume which is injected is 20 μL and of the matrix modifier 5 μL.

| Parameters Pb | Settings |
|---|---|
| lamp current | 5 mA |
| Wavelength | 283.3 nm |
| slit width | 0.5 nm |
| Measurement mode | peak area |
| Replicates | 3 |

| Temperature program | | | |
|---|---|---|---|
| Step | Temp (°C) | Time (s) | Sheath gas |
| 1 | 95 | 5.0 | $Ar/H_2$ |
| 2 | 130 | 40.0 | $Ar/H_2$ |
| 3 | 800 | 5.0 | $Ar/H_2$ |
| 4 | 800 | 5.0 | $Ar/H_2$ |
| 5 | 800 | 2.0 | Ar |
| 6 | 2100 | 0.7 | - |
| 7 | 2100 | 2.0 | - |
| 8 | 2500 | 0.3 | - |
| 9 | 2500 | 3.0 | Ar |

Ar = argon, $Ar/H_2$ = 95 % argon and 5 % hydrogen

The temperature program given here is for the digestion procedure 2.4 ($HNO_3$ - HF - $H_2O_2$). For other digestion procedures the temperature program should be optimised.

7. When using other instruments or matrix modifiers, the optimum temperature values to be set may differ from the values given above. The temperatures mentioned are instrument settings instead of real temperature values; such settings may differ even within two instruments of the same type and should be always checked out experimentally.

# 9. CALCULATION

9.1 The total lead content in the dried plant material, expressed in μg/kg Pb, is calculated by:

(a - b) * V / w

in which:

a is the concentration of lead in the sample digest, in μg/L;

b is the concentration of lead in the blank digest, in μg/L;

V is the total volume of digest at the end of the digestion procedure, in mL;

w is the weight of plant material sample, in g.

# 10. REFERENCES

10.1 Van der Lee, J.J., E.J.M. Temminghoff, V.J.G. Houba and I. Novozamsky. 1987. Background corrections in the determination of Cd and Pb by flame AAS in plant and soil samples with high Fe levels. Appl. Spectrosc. 41: 388-390.

10.2 Temminghoff, E.J.M. 1990. Signal stabilisation in Electrothermal Atomisation Atomic Absorption Spectrometry by means of addition of butanol. J. Anal. At. Spectrom. 5: 273.

# 4.18.D DETERMINATION OF LEAD BY ICP-MS

## 1. PRINCIPLE OF THE METHOD

1.1 Solutions with lead compounds are nebulised into an argon plasma, where all components are vaporised. The ions produced are entrained in the plasma gas and introduced into a mass spectrometer, sorted according to their mass-to-charge ratios, and quantified with a channel electron multiplier. Lead is determined by the sum of mass 206, 207, and 208 amu.

1.2 This determination may be carried out on digest 2.3 (microwave digestion with $HNO_3$ - $H_2O_2$ - HF), digest 2.4 (digestion with $HNO_3$ - $H_2O_2$ - HF) and digest 2.7 (digestion by dry-ashing followed by treatment with HF).

## 2. RANGE AND DETECTION LIMIT

2.1 This procedure yields a standard curve that is linear up to approximately 50 µg/L Pb.

2.2 The detection limit is approximately 0.04 µg/L in the digest. The determination limit is approximately 0.12 µg/L (13 respectively 40 µg/kg in the dried plant material).

## 3. INTERFERENCES

3.1 No interferences are expected.

## 4. PRECISION AND ACCURACY

4.1 The reproducibility of determinations by this procedure should give, at reasonable control and thorough sample preparation, a coefficient of variation within 10 %.

## 5. APPARATUS

5.1 Inductively coupled plasma mass spectrometer.

## 6. REAGENTS

6.1A Stock Solution, Pb concentration 1000 mg/L - Merck nr 1.02607.

6.1B Stock Solution, Pb concentration 1000 mg/L - Dissolve 1.5985 g lead nitrate, $Pb(NO_3)_2$, in some ultra pure water in a volumetric flask of 1000 mL, Make up to 1000 mL with ultra pure water.

6.2 Standard Solution, Pb concentration 1 mg/L - Pipette 1 mL concentrated nitric acid (65 % s.p.) in a 1000-mL polythene volumetric flask which already contains about

500 mL ultra pure water. Add 1.00 mL stock solution (6.1A or 6.1B) and make up to volume with ultra pure water.

Remark:

1. $Pb(NO_3)_2$ may absorb water on standing. The reagent should be standardised by titration with EDTA at pH=10 in the presence of tartrate ions with Eriochrome Black T as an indicator.

## 7. CALIBRATION AND STANDARDS

7.1 Standard Series - pipette 0 – 1.00 – 2.00 – 5.00 mL of the standard solution (6.2) in about 40 mL ultra pure water in a 100-mL polythene volumetric flasks. Add either 1.0 mL concentrated nitric acid (65 %) (digestion 2.3), 0.3 mL concentrated nitric acid (65 %) (digestion 2.4) or 0.1 mL concentrated hydrochloric acid (36 %) (digestion 2.7). Let cool down and make up to the mark with ultra pure water. This standard series has Pb concentrations of 0 –10 – 20 – 50 µg/L.

7.2 Calibration Curve - The counts per second are plot versus µg/L lead in the standard series.

Remark:

2. A mix standard series can be used for simultaneous measurements with ICP-MS (Al, As, Cd, Co, Cr, Cu, Mn, Ni, Pb, Sb, Sn, V and Zn).

## 8. PROCEDURE

8.1 Measurement – Measure in the standard series, the blanks and the sample digests the Pb concentration with the ICP-MS by the sum of mass 206, 207, and 208 amu.

Remarks:

3. A data management system and system controller are used. In this way the measurements are checked continuously and the data output is in concentration units in the extracts.
4. Due to the linearity of the ICP-MS it is possible to calibrate the ICP-MS only with the highest and zero standard of the standard series. As a check all standards can be measured as samples.

## 9. CALCULATION

9.1 The total lead content in the dried plant material, expressed in µg/kg Pb, is calculated by:

$$(a - b) * V / w$$

in which:

a is the concentration of lead in the sample digest, in µg/L;

b is the concentration of lead in the blank digest, in µg/L;

V is the total volume of digest at the end of the digestion procedure, in mL;

w is the weight of plant material sample, in g.

# 4.19 DETERMINATION OF SULPHUR FRACTIONS

## 4.19.A DETERMINATION OF TOTAL SULPHUR BY ICP-OES

### 1. PRINCIPLE OF THE METHOD

1.1 Solutions with sulphur compounds are nebulised into an argon plasma, where all components are vaporised. Sulphur compounds are dissociated and excited, and then emit radiation of which the intensity is measured at a wavelength of 181.972 nm.

1.2 This determination may be carried out on digest 2.3 (microwave digestion with $HNO_3$ - $H_2O_2$ - HF), digest 2.4 (digestion with $HNO_3$ - $H_2O_2$ - HF) and digest 2.5 (digestion with $HNO_3$).

### 2. RANGE AND DETECTION LIMIT

2.1 This procedure yields a standard curve that is linear up to at least 80 mg/L S.

2.2 The detection limit is approximately 0.05 mg/L in the digest. The determination limit is approximately 0.14 mg/L (0.5 respectively 1.5 mmol/kg in the dried plant material).

### 3. INTERFERENCES

3.1 No interferences are expected.

### 4. PRECISION AND ACCURACY

4.1 The reproducibility of determinations by this procedure should give, at reasonable control and thorough sample preparation, a coefficient of variation within 10 %.

### 5. APPARATUS

5.1 Inductively coupled plasma optical emission spectrometer.

### 6. REAGENTS

6.1A <u>Stock Solution, S concentration 1000 mg/L</u> – Merck nr 1.09872. Transfer 3 ampoules of 1000 mg/L $SO_4^{2-}$ a 1000-mL volumetric flask and make up to volume with water.

6.1B <u>Stock Solution, S concentration 1000 mg/L</u> - Dissolve 5.4398 g potassium sulphate, $K_2SO_4$ (see remark 1), in some water in a 1000-mL volumetric flask and make up to volume.

Remark:

1. The potassium sulphate has to be dried at 200 °C for at least 24 hours just before weighing. Just before drying, any lumps should be cut down so that only fine crystals remain. This is important, because coarse crystals contain much occluded water in their cavities and thus would need a higher drying temperature. The fine crystals should, however, not be pulverised because this enhances deliquescence (= attraction of water) and/or efflorescence (= weathering) later on. After the recommended heating procedure, the potassium sulphate should be cooled in a desiccator containing preferably magnesium perchlorate as a desiccant. The dried potassium sulphate should be weighed as soon as it has reached the ambient temperature.

## 7. CALIBRATION AND STANDARDS

7.1 Standard Series - Pipette 0 – 1.00 – 2.00 – 8.00 mL of the stock solution (6.1A or 6.1B) into 100-mL volumetric flasks which already contain 40 mL water. Add either 10 mL concentrated nitric acid (65 %) (digestion 2.3), 3.0 mL concentrated nitric acid (65 %) (digestion 2.4) or 1.5 mL concentrated nitric acid (65 %) (digestion 2.5). Allow cooling and make up to volume. This standard series has S concentrations of 0 – 10 – 20 – 80 mg/L.

7.2 Calibration Curve - The emission counts are plot versus mg/L sulphur in the standard series.

Remark:

2. A mix standard series can be used for simultaneous measurements with ICP-OES (Al, Ca, Cd, Co, Cr, Cu, Fe, K, Mg, Mn, Na, P, S and Zn).

## 8. PROCEDURE

8.1 Measurement - Measure in the standard series, the blanks and sample digests the S concentration with an ICP-OES at a wavelength of 181.972 nm. At this wavelength a (fitted) background correction is used.

Remarks:

3. A data management system and system controller are used. In this way the measurements are checked continuously and the data output is in concentration units in the extracts.
4. Due to the linearity of the ICP-OES it is possible to calibrate the ICP-OES only with the highest and zero standard of the standard series. As a check all standards can be measured as samples.
5. Beryllium (5 mg/L), at a wavelength of 234,861 nm, is used in the author's laboratory as an internal standard to compensate for matrix effects.

## 9. CALCULATION

9.1 The total sulphur content of the dried plant material, expressed in mmol/kg S, is calculated by:

$0.03119 * (a - b) * V / w$

in which:
a is the concentration of sulphur in the sample digest, in mg/L;
b is the concentration of sulphur in the blank digest, in mg/L;
V is the total volume of digest at the end of the digestion procedure, in mL;
w is the weight of plant material sample, in g.

## 10. REFERENCES

10.1 Novozamsky, I., R. van Eck, J.J. van der Lee, V.J.G. Houba and E.J.M. Temminghoff. 1986. Determination of total sulphur and extractable sulphate in plant materials by inductively-coupled plasma atomic emission spectrometry. Commun. Soil Sci. Plant Anal. 17: 1147-1157.

# 4.19.B DETERMINATION OF SULPHATE BY ICP-OES

## 1. PRINCIPLE OF THE METHOD

1.1 Sulphate ions in the extracts are precipitated by barium ions. After a clean up, the barium sulphate precipitate is dissolved by EDTA. The solution is then nebulised into an argon plasma, where all components are vaporised. Sulphate ions decompose; the sulphur atoms thus formed are excited and then emit radiation of which the intensity is measured at a wavelength of 181.972 nm.

1.2 This determination may be carried out on extract 3.1 (extraction with water).

## 2. RANGE AND DETECTION LIMIT

2.1 This procedure yields a standard curve that is linear up to at least 80 mg/L S.

2.2 The detection limit is approximately 0.05 mg/L in the extract. The determination limit is approximately 0.15 mg/L (0.5 respectively 1.5 mmol/kg in the dried plant material).

## 3. INTERFERENCES

3.1 No interferences are expected.

## 4. PRECISION AND ACCURACY

4.1 The reproducibility of determinations by this procedure should give, at reasonable control and thorough sample preparation, a coefficient of variation within 10 % + 10 mmol/kg.

## 5. APPARATUS

5.1 Inductively coupled plasma optical emission spectrometer.

5.2 Centrifuge with swing-out head.

## 6. REAGENTS

6.1A Stock Solution, S concentration 1000 mg/L - Merck nr 1.09872. Transfer 3 ampoules of 1000 mg/L $SO_4^{2-}$ to a 1000-mL volumetric flask and make up to volume with water.

6.1B Stock Solution, S concentration 1000 mg/L - Dissolve 5.4398 g potassium sulphate, $K_2SO_4$ (see remark 1), in some water in a 1000-mL volumetric flask and make up to volume.

6.2 Standard Solution, S concentration 400 mg/L - Pipette 40 mL stock solution (6.1A or 6.1B) in a 100-mL volumetric flask and make up to volume.

6.3 Barium Chloride Solution 1 mol/L - Dissolve 244 g barium chloride dihydrate, $BaCl_2.2H_2O$, in about 600 mL water. Add 82 mL concentrated hydrochloric acid (36 %) and make up to 1 litre.

6.2 EDTA Solution 0.02 mol/L - Dissolve 5.84 g ethylene diamine tetra acetic acid, $H_4EDTA$, in 30 mL concentrated aqueous ammonia, $NH_4OH$ ($\rho = 0.91$ g/cm$^3$). Make up to 1 litre with water.

Remark:

1. The potassium sulphate has to be dried at 200 °C for at least 24 hours just before weighing. Just before drying, any lumps should be cut down so that only fine crystals remain. This is important, because coarse crystals contain much occluded water in their cavities and thus would need a higher drying temperature. The fine crystals should, however, not be pulverised because this enhances deliquescence (= attraction of water) and/or efflorescence (= weathering) later on. After the recommended heating procedure, the potassium sulphate should be cooled in a desiccator containing preferably magnesium perchlorate as a desiccant. The dried potassium sulphate should be weighed as soon as it has reached the ambient temperature.

## 7. CALIBRATION AND STANDARDS

7.1 Standard Series - Pipette 0 – 1.00 – 2.00 – 8.00 mL of the stock solution (6.1A or 6.1B) into 100-mL volumetric flasks. Make up to the mark with EDTA solution (6.3). This standard series has S concentrations of 0 – 10 – 20 – 80 mg/L.

7.2 Calibration Curve - The emission counts are plot versus mg/L sulphur in the standard series.

## 8. PROCEDURE

8.1 Separation of sulphate - Weigh a series of 15-mL centrifuge tubes (empty weight e gram per tube). Add to each tube 1.0 mL of barium chloride solution (6.3). Then pipette 5.0 mL of the standard series, the blanks and the sample extracts into these tubes, and mix by shaking. Bring the tubes two by two with water at the same weight, and centrifuge for 10 min at 1200-1500 g. Carefully decant the supernatant, so that the precipitate and only about 0.5 mL of liquid will remain in the tube. Next, add 5.0 mL of water and repeat the shaking and the centrifuging. Decant again the supernatant and weigh the tube (c gram) in order to establish the volume of liquid left behind. Finally, add 5.0 mL of EDTA solution (6.4) and shake to bring the precipitate into suspension. Allow the precipitate to dissolve completely, preferably overnight but at least during 2 hours; swirl now and then.

8.2 Measurement - Measure in the standard series, the blanks and the sample extracts the sulphate-sulphur concentration with an ICP-OES at a wavelength of 181.972 nm. At this wavelength a (fitted) background correction is used.

Remarks:

2. To measure low concentrations of sulphate sulphur in the sample extracts a 1:1 dilution of the extract with EDTA can be used instead of the 1:5 dilution. The calculation should be amended accordingly.
3. A data management system and system controller are used. In this way the measurements are checked continuously and the data output is in concentration units in the extracts.
4. Due to the linearity of the ICP-OES it is possible to calibrate the ICP-OES only with the highest and zero standard of the standard series. As a check all standards can be measured as samples.
5. Beryllium (5 mg/L), at a wavelength of 234,861 nm, is used in the author's laboratory as an internal standard to compensate for matrix effects.

## 9. CALCULATION

9.1 The sulphate-sulphur content in the dried plant material, expressed in mmol/kg S-$SO_4$, is calculated by:

(5 + c - e) / 5 * 0.03119 * (a - b) * V / w

in which:

c is weight of centrifuge tube with remaining liquid, in g;

e is weight of empty centrifuge tube, in g;

a is the concentration of sulphate-sulphur in the plant material digest, in mg/L;

b is the concentration of sulphate-sulphur in the sample extract, in mg/L;

V is the total volume of water taken for extraction, in mL;

w is the weight of plant material sample, in g.

## 10. REFERENCES

10.1 Novozamsky, I., R. van Eck, J.J. van der Lee, V.J.G. Houba and E.J.M. Temminghoff. 1986. Determination of total sulphur and extractable sulphate in plant materials by inductively coupled plasma atomic emission spectrometry. Commun. Soil Sci. Plant Anal. 17: 1147-1157.

# 4.20 DETERMINATION OF ANTIMONY

## 4.20.A DETERMINATION OF ANTIMONY BY ICP-MS

### 1. PRINCIPLE OF THE METHOD

1.1 Solutions with antimony compounds are nebulised into an argon plasma, where all components are vaporised. The ions produced are entrained in the plasma gas and introduced into a mass spectrometer, sorted according to their mass-to-charge ratios, and quantified with a channel electron multiplier. Antimony is determined at mass 121 amu.

1.2 This determination may be carried out on digest 2.3 (microwave digestion with $HNO_3$ - $H_2O_2$ - HF), digest 2.4 (digestion with $HNO_3$ - $H_2O_2$ - HF) and digest 2.7 (digestion by dry-ashing followed by treatment with HF).

### 2. RANGE AND DETECTION LIMIT

2.1 This procedure yields a standard curve that is linear up to approximately 50 μg/L Sb.

2.2 The detection limit is approximately 0.005 μg/L in the digest. The determination limit is approximately 0.015 μg/L (2 respectively 5 μg/kg in the dried plant material).

### 3. INTERFERENCES

3.1 No interferences are expected

### 4. PRECISION AND ACCURACY

4.1 The reproducibility of determinations by this procedure should give, at reasonable control and thorough sample preparation, a coefficient of variation within 10 %.

### 5. APPARATUS

5.1 Inductively coupled plasma mass spectrometer.

### 6. REAGENTS

6.1 <u>Stock Solution, Sb concentration 1000 mg/L</u> - Merck nr 1.02601.

6.2 <u>Standard Solution, Sb concentration 1 mg/L</u> - Pipette 1 mL concentrated nitric acid (65 % s.p.) in a 1000-mL polythene volumetric flask which already contains about

500 mL ultra pure water. Add 1.00 mL stock solution (6.1) and make up to volume with ultra pure water.

## 7. CALIBRATION AND STANDARDS

7.1 Standard Series - pipette 0 – 1.00 – 2.00 – 5.00 mL of the standard solution (6.2) in about 40 mL ultra pure water in a 100-mL polythene volumetric flasks. Add either 1.0 mL concentrated nitric acid (65 %) (digestion 2.3), 0.3 mL concentrated nitric acid (65 %) (digestion 2.4) or 0.1 mL concentrated hydrochloric acid (36 %) (digestion 2.7). Let cool down and make up to the mark with ultra pure water. This standard series has Sb concentrations of 0 – 10 – 20 – 50 µg/L.

7.2 Calibration Curve - The counts per second are plot versus µg/L antimony in the standard series.

Remark:

1. A mix standard series can be used for simultaneous measurements with ICP-MS (Al, As, Cd, Co, Cr, Cu, Mn, Ni, Pb, Sb, Sn, V and Zn).

## 8. PROCEDURE

8.1 Measurement - Measure in the standard series, the blanks and the sample digests the Sb concentration with the ICP-MS at a mass of 121 amu.

Remarks:

2. A data management system and system controller are used. In this way the measurements are checked continuously and the data output is in concentration units in the digests.
3. Due to the linearity of the ICP-MS it is possible to calibrate the ICP-MS only with the highest and zero standard of the standard series. As a check all standards can be measured as samples.

## 9. CALCULATION

9.1 The total antimony content in the dried plant material, expressed in µg/kg Sb, is calculated by:

$(a - b) * V / w$

in which:

a is the concentration of antimony in the sample digest, in µg/L;
b is the concentration of antimony in the blank digest, in µg/L;
V is the total volume of digest at the end of the digestion procedure, in mL;
w is the weight of plant material sample, in g.

# 4.21 DETERMINATION OF SILICON

## 4.21.A DETERMINATION OF SILICON BY ICP-OES

### 1. PRINCIPLE OF THE METHOD

1.1 Solutions with silicon compounds are nebulised into an argon plasma, where all components are vaporised. Silicon compounds are dissociated and excited, and then emit radiation of which the intensity is measured at a wavelength of 288.158 nm.

1.2 This determination is to be carried out on extract 3.2 (extraction with HF - HCl). The silicon content of the dried plant material should be higher than 500 mg/kg.

### 2. RANGE AND DETECTION LIMIT

2.1 This procedure yields a standard curve that is linear up to at least 100 mg/L Si.

2.2 The detection limit is approximately 0.5 mg/L in the extract. The determination limit is approximately 1.5 mg/L (16 respectively 50 mg/kg in the dried plant material).

### 3. INTERFERENCES

3.1 No interferences are expected.

Remark:
1. Only polythene flasks etc. can be used because glass contains Si.

### 4. PRECISION AND ACCURACY

4.1 The reproducibility of determinations by this procedure should give, at reasonable control and thorough sample preparation, a coefficient of variation within 10 % + 10 mg/kg.

### 5. APPARATUS

5.1 Polycarbonate test tubes + stoppers.

5.2 Polythene bottles.

5.3 Polythene volumetric flasks.

5.3 Inductively coupled plasma optical emission spectrometer, fitted with HF-resistant nebuliser and inner torch tube.

# 6. REAGENTS

6.1 Stock Solution, Si concentration 1000 mg/L - LPS Benelux art. nr 3740.1.

6.2 Acid mixture - Mix 150 mL of concentrated hydrofluoric acid (40 %) with 60 mL of concentrated hydrochloric acid (36 %) and add 165 mL ultra pure water. (This solution is 9.0 M in HF and 2.0 M in HCl.

6.3 Ammonia Solution 4 mol/L - Dilute 75 mL of concentrated aqueous ammonia (25 %) with water to 250 mL. See remark 5.

Remarks:

2. For this determination, all solutions must be prepared with ultra pure water, since demineralised water may contain varying amounts of Si.
3. A mix standard series can be used for simultaneous measurements with ICP-OES (B and Si).

# 7. CALIBRATION AND STANDARDS

7.1 Standard Series - Pipette 0 – 5.00 – 10.00 mL stock solution (6.1) in 100-mL polythene volumetric flasks, add 50 ml acid mixture (6.2) and make up to volume with ultra pure water. This standard series has Si concentrations of 0 – 50 – 100 mg/L.

7.2 Calibration Curve - The emission counts are plot versus mg/L silicium in the standard series.

Remarks:

4. The dilution for the standards and samples must be done with ultra pure water of one batch, i.e., having the same quality.
5. Hydrofluoric acid is a treacherous skin poison. Work in a good fume hood and wear rubber gloves, safety goggles and protective cloths. If any HF comes into contact with the skin, wash immediately and thoroughly with water and thereafter dab with 4 M ammonia (4.4) or calcium gluconate gel.

# 8. PROCEDURE

8.1 Measurement - Dilute, using piston-type pipettes or a diluting apparatus with polythene tips the standard series, the blanks and the sample extracts 1 + 19 (v/v) with ultra pure water in polycarbonate test tubes and mix. Measure in the standard series, the blanks and the sample extracts the Si concentration with the ICP-OES at a wavelength of 251.61 nm.

Remarks:

5. New polycarbonate test tubes, pipette tips etc. must be cleaned by rinsing with the acid mixture (6.2), because new material may release silicon.
6. Samples with Si concentrations beyond the standard series should be diluted first with acid mixture (6.2) and then follow the procedure with the usual 1:20 dilution.
7. Beryllium (5 mg/L), at a wavelength of 234,861 nm, is used in the author's laboratory as an internal standard to compensate for matrix effects.

## 9. CALCULATION

9.1 The silicon content of the dried plant material, expressed in mg/kg Si, is calculated by:

$(a - b) * V / w$

in which:

a is the concentration of silicon in the sample extract, in mg/L;

b is the concentration of silicon in the blank extract, in mg/L;

V is the volume of acid mixture used for extraction, in mL;

w is the weight of plant material sample, in g.

## 10. REFERENCES

10.1 Novozamsky, I., R. van Eck and V.J.G. Houba. 1984. A rapid determination of silicon in plant material. Commun. Soil Sci. Plant Anal., 15: 205-211.

# 4.22 DETERMINATION OF TIN

# 4.22.A DETERMINATION OF TIN BY ICP-MS

## 1. PRINCIPLE OF THE METHOD

1.1 Solutions with tin compounds are nebulised into an argon plasma, where all components are vaporised. The ions produced are entrained in the plasma gas and introduced into a mass spectrometer, sorted according to their mass-to-charge ratios and quantified with a channel electron multiplier. Tin is determined at mass 120 amu.

1.2 This determination may be carried out on digest 2.3 (microwave digestion with $HNO_3$ - $H_2O_2$ - HF), digest 2.4 (digestion with $HNO_3$ - $H_2O_2$ - HF) and digest 2.7 (digestion by dry-ashing followed by treatment with HF).

## 2. RANGE AND DETECTION LIMIT

2.1 This procedure yields a standard curve that is linear up to at least 50 μg/L Sn.

2.2 The detection limit is approximately 0.08 μg/L in the digest. The determination limit is approximately 0.23 μg/L (27 respectively 80 μg/kg in the dried plant material).

## 3. INTERFERENCES

3.1 $^{120}$Te has the same mass as $^{120}$Sn. A correction factor can be applied. $^{78}$Se$^{40}$Ar molecule has the same mass as $^{118}$Sn whereas $^{80}$Se$^{40}$Ar has the same mass as $^{120}$Sn.

## 4. PRECISION AND ACCURACY

4.1 The reproducibility of determinations by this procedure should give, at reasonable control and thorough sample preparation, a coefficient of variation within 10 %.

## 5. APPARATUS

5.1 Inductively coupled plasma mass spectrometer.

## 6. REAGENTS

6.1 Stock Solution, Sn concentration 1000 mg/L - Merck nr 1.02671.

6.2 Standard Solution, Sn concentration 1 mg/L - Pipette 1 mL concentrated nitric acid (65 % s.p.) in a 1000-mL polythene volumetric flask which already contains about 500 mL ultra pure water. Add 1.00 mL stock solution (6.1) and make up to volume with ultra pure water.

# 7. CALIBRATION AND STANDARDS

7.1 Standard Series - pipette 0 – 1.00 – 2.00 – 5.00 mL of the standard solution (6.2) in about 40 mL ultra pure water in a 100-mL polythene volumetric flasks. Add either 1.0 mL concentrated nitric acid (65 %) (digestion 2.3), 0.3 mL concentrated nitric acid (65 %) (digestion 2.4) or 0.1 mL concentrated hydrochloric acid (36 %) (digestion 2.7). Let cool down and make up to the mark with ultra pure water. This standard series has Sn concentrations of 0 –10 – 20 – 50 µg/L.

7.2 Calibration Curve - The counts per second are plot versus µg/L tin in the standard series.

Remark:

1. A mix standard series can be used for simultaneous measurements with ICP-MS (Al, As, Cd, Co, Cr, Cu, Mn, Ni, Pb, Sb, Sn, V and Zn).

# 8. PROCEDURE

8.1 Measurement – Measure in the standard series, the blanks and the sample digests the Sn concentration with the ICP-MS at a mass of 120 amu.

Remarks:

2. A data management system and system controller are used. In this way the measurements are checked continuously and the data output is in concentration units in the digests.
3. Due to the linearity of the ICP-MS it is possible to calibrate the ICP-MS only with the highest and zero standard of the standard series. As a check all standards can be measured as samples.

# 9. CALCULATION

9.1 The total tin content in the dried plant material, expressed in µg/kg Sn, is calculated by:

$(a - b) * V / w$

in which:

a is the concentration of tin in the sample digest, in µg/L;

b is the concentration of tin in the blank digest, in µg/L;

V is the total volume of digest at the end of the digestion procedure, in mL;

w is the weight of plant material sample, in g.

# 4.23 DETERMINATION OF VANADIUM

## 4.23.A DETERMINATION OF VANADIUM BY ICP-MS

### 1. PRINCIPLE OF THE METHOD

1.1 Solutions with vanadium compounds are nebulised into an argon plasma, where all components are vaporised. The ions produced are entrained in the plasma gas and introduced into a mass spectrometer, sorted according to their mass-to-charge ratios and quantified with a channel electron multiplier. Vanadium is determined at mass 51 amu.

1.2 This determination may be carried out on digest 2.3 (microwave digestion with $HNO_3$ - $H_2O_2$ - HF), digest 2.4 (digestion with $HNO_3$ - $H_2O_2$ - HF) and digest 2.7 (digestion by dry-ashing followed by treatment with HF).

### 2. RANGE AND DETECTION LIMIT

2.1 This procedure yields a standard curve that is linear up to at least 50 μg/L V.

2.2 The detection limit is approximately 0.005 μg/L in the digest. The determination limit is approximately 0.015 μg/L (2 respectively 5 μg/kg in the dried plant material).

### 3. INTERFERENCES

3.1 Interferences are expected for $^{34}S^{16}O^{1}H$ and $^{35}Cl^{16}O$ due to mass overlap with $^{51}V$.

### 4. PRECISION AND ACCURACY

4.1 The reproducibility of determinations by this procedure should give, at reasonable control and thorough sample preparation, a coefficient of variation within 10 %.

### 5. APPARATUS

5.1 Inductively coupled plasma mass spectrometer.

### 6. REAGENTS

6.1 Stock Solution, V concentration 1000 mg/L - Merck nr 1.02666.

6.2 Standard Solution, V concentration 1 mg/L - Pipette 1 mL concentrated nitric acid (65 % s.p.) in a 1000-mL polythene volumetric flask which already contains about 500 mL ultra pure water. Add 1.00 mL stock solution (6.1) and make up to volume with ultra pure water.

# 7. CALIBRATION AND STANDARDS

7.1 Standard Series - pipette 0 – 1.00 – 2.00 – 5.00 mL of the standard solution (6.2) in about 40 mL ultra pure water in a 100-mL polythene volumetric flasks. Add either 1.0 mL concentrated nitric acid (65 %) (digestion 2.3), 0.3 mL concentrated nitric acid (65 %) (digestion 2.4) or 0.1 mL concentrated hydrochloric acid (36 %) (digestion 2.7). Let cool down and make up to the mark with ultra pure water. This standard series has V concentrations of 0 –10 – 20 – 50 μg/L.

7.2 Calibration Curve - The counts per second are plot versus μg/L vanadium in the standard series.

Remark:

1. A mix standard series can be used for simultaneous measurements with ICP-MS (Al, As, Cd, Co, Cr, Cu, Mn, Ni, Pb, Sb, Sn, V and Zn).

# 8. PROCEDURE

8.1 Measurement - Measure in the standard series, the blanks and the sample digests the V concentration with the ICP-MS at a mass of 51 amu. Make use of corrections if necessary.

Remarks:

2. A data management system and system controller are used. In this way the measurements are checked continuously and the data output is in concentration units in the digests.
3. Due to the linearity of the ICP-MS it is possible to calibrate the ICP-MS only with the highest and zero standard of the standard series. As a check all standards can be measured as samples.

# 9. CALCULATION

9.1 The total vanadium content in the dried plant material, expressed in μg/kg V, is calculated by:

(a - b) * V / w

in which:

a is the concentration of vanadium in the sample digest, in μg/L;

b is the concentration of vanadium in the blank digest, in μg/L;

V is the total volume of digest at the end of the digestion procedure, in mL;

w is the weight of plant material sample, in g.

# 4.24 DETERMINATION OF ZINC

# 4.24.A DETERMINATION OF ZINC BY FLAME AAS

## 1. PRINCIPLE OF THE METHOD

1.1 The sample is nebulised into an air-acetylene flame, where it is vaporised; zinc compounds are atomised and the zinc atoms thus formed absorb radiation from a hollow cathode lamp. The absorbance is measured at a wavelength of 213.9 nm, using deuterium background correction.

1.2 This determination may be carried out on digest 2.1 (digestion with $H_2SO_4$ - salicylic acid - $H_2O_2$ - Se), digest 2.2 (digestion with $H_2SO_4$ - salicylic acid - $H_2O_2$), digest 2.3 (microwave digestion with $HNO_3$ - $H_2O_2$ - HF), digest 2.4 (digestion with $HNO_3$ - $H_2O_2$ - HF), digest 2.6 (digestion with $HNO_3$ - $HClO_4$ - $H_2SO_4$) and digest 2.7 (digestion by dry-ashing followed by treatment with HF).

## 2. RANGE AND DETECTION LIMIT

2.1 This procedure yields a standard curve that is linear to approximately 0.8 mg/L Zn.

2.2 The detection limit is approximately 0.02 mg/L in the digest. The determination limit is approximately 0.06 mg/L (3.0-7.5 mg/kg in the dried plant material).

## 3. INTERFERENCES

3.1 To avoid interferences as much as possible, background correction is necessary.

## 4. PRECISION AND ACCURACY

4.1 The reproducibility of determinations by this procedure should give, at reasonable control and thorough sample preparation, a coefficient of variation within 5 % + 5 mg/kg.

## 5. APPARATUS

5.1 Flame atomic absorption spectrophotometer with a device for correcting background absorption.

## 6. REAGENTS

6.1A <u>Stock Solution, Zn concentration 1000 mg/L</u> - Merck nr 1.19806.

6.1B <u>Stock Solution, Zn concentration 1000 mg/L</u> - Dissolve 4.398 g zinc sulphate heptahydrate, $ZnSO_4.7H_2O$, in about 500 mL ultra pure water in a 1000-mL volumetric flask and make up to volume.

6.2 <u>Standard Solution, Zn concentration 20 mg/L</u> - Pipette 2.00 mL stock solution (6.1A or 6.1B) into a 100-mL volumetric flask and make up to volume.

<u>Remark</u>:

1. $ZnSO_4.7H_2O$ may lose crystal water on standing. The reagent should be standardised by titration with EDTA at pH 10 with Eriochrome Black T as an indicator.

## 7. CALIBRATION AND STANDARDS

7.1 <u>Standard Series</u> - Pipette 0 – 2.00 – 4.00 – 6.00 – 8.00 – 10.00 mL of the standard solution (6.2) into 100-mL volumetric flasks which already contain 40 mL water. Add either 4.5 mL concentrated sulphuric acid (96 %) (digestion 2.1 or 2.2), 10 mL concentrated nitric acid (65 %) (digestion 2.3), 3.0 mL concentrated nitric acid (65 %) (digestion 2.4), 0.45 mL concentrated sulphuric acid (96 %) (digestion 2.6) or 1.0 mL concentrated hydrochloric acid (36 %) (digestion 2.7). Let cool down and make up to the mark with water. This standard series has Zn concentrations of 0 - 0.2 – 0.4 – 0.8 – 1.2 – 1.6 – 2.0 mg/L.

7.2 <u>Calibration Curve</u> - The absorbance (A) is plot versus mg/L zinc in the standard series.

<u>Remark</u>:

2. The calibration curve is slightly bent towards the x-axis. This means that calculation by means of linear regression is not allowed.

## 8. PROCEDURE

8.1 <u>Measurement</u> – Measure in the standard series, the blanks and the sample digests the Zn concentration with flame AAS at a wavelength of 213.9 nm, using a just blue (stoichiometric) air-acetylene flame. Use background correction.

<u>Remark</u>:

3. Do not use rubber stoppers, since these may release zinc.

## 9. CALCULATION

9.1 The total zinc content in the dried plant material, expressed in mg/kg Zn, is calculated by:

(a - b) * V / w

in which:

a is the concentration of zinc in the sample digest, in mg/L;
b is the concentration of zinc in the blank digest, in mg/L;
V is the total volume of digest at the end of the digestion procedure, in mL;
w is the weight of plant material sample, in g.

# 4.24.B DETERMINATION OF ZINC BY ICP-OES

## 1. PRINCIPLE OF THE METHOD

1.1 Solutions with zinc compounds are nebulised into an argon plasma, where all components are vaporised. Zinc compounds are dissociated and excited, and then emit radiation of which the intensity is measured at a wavelength of 206.200 nm.

1.2 This determination may be carried out on digest 2.3 (microwave digestion with $HNO_3$ - $H_2O_2$ - HF), digest 2.4 (digestion with $HNO_3$ - $H_2O_2$ - HF) and digest 2.7 (digestion by dry-ashing followed by treatment with HF).

## 2. RANGE AND DETECTION LIMIT

2.1 This procedure yields a standard curve that is linear up to approximately 40 mg/L Zn.

2.2 The detection limit is approximately 0.02 mg/L in the digest. The determination limit is approximately 0.06 mg/L (3.0-7.5 mg/kg in the dried plant material).

## 3. INTERFERENCES

3.1 No interferences are expected.

## 4. PRECISION AND ACCURACY

4.1 The reproducibility of determinations by this procedure should give, at reasonable control and thorough sample preparation, a coefficient of variation within 5 % + 5 mg/kg.

## 5. APPARATUS

5.1 Inductively coupled plasma optical emission spectrometer.

## 6. REAGENTS

6.1A Stock Solution, Zn concentration 1000 mg/L - Merck nr 1.19806.

6.1B Stock Solution, Zn concentration 1000 mg/L - Dissolve 4.398 g zinc sulphate heptahydrate, $ZnSO_4.7H_2O$, in about 500 mL water in a 1000-mL volumetric flask and make up to volume.

Remark:

1. $ZnSO_4.7H_2O$ may lose crystal water on standing. The reagent should be standardised by titration with EDTA at pH 10 with Eriochrome Black T as an indicator.

# 7. CALIBRATION AND STANDARDS

7.1 Standard Series - Pipette 0 – 0.10 – 0.20 – 4.00 mL of the stock solution (6.1A or 6.1B) into 100-mL volumetric flasks which already contain 40 mL water. Add either 10 mL concentrated nitric acid (65 %) (digestion 2.3), 3.0 mL concentrated nitric acid (65 %) (digestion 2.4) or 1.0 mL concentrated hydrochloric acid (36 %) (digestion 2.7). Let cool down and make up to the mark with water. This standard series has Zn concentrations of 0 – 1.0 – 2.0 – 40.0 mg/L.

7.2 Calibration Curve - The emission counts are plot versus mg/L zinc in the standard series.

Remark:

2. A mix standard series can be used for simultaneous measurements with ICP-OES (Al, Ca, Cd, Co, Cr, Cu, Fe, K, Mg, Mn, Na, Ni, P, S and Zn).

# 8. PROCEDURE

8.1 Measurement - Measure in the standard series, the blanks and the sample digests the Zn concentration with the ICP-OES at a wavelength of 213.856 nm. At this wavelength a (fitted) background correction is used.

Remarks:

3. A data management system and system controller are used. In this way the measurements are checked continuously and the data output is in concentration units in the digests.
4. Due to the linearity of the ICP-OES it is possible to calibrate the ICP-OES only with the highest and zero standard of the standard series. As a check all standards can be measured as samples.
5. Scandium (5 mg/L), at a wavelength of 255.235 nm, is used in the author's laboratory as an internal standard to compensate for matrix effects.

# 9. CALCULATION

9.1 The total zinc content in the dried plant material, expressed in mg/kg Zn, is calculated by:

$(a - b) * V / w$

in which:

a is the concentration of zinc in the sample digest, in mg/L;

b is the concentration of zinc in the blank digest, in mg/L;

V is the total volume of digest at the end of the digestion procedure, in mL;

w is the weight of plant material sample, in g.

# 4.24.C DETERMINATION OF ZINC BY ICP-MS

## 1. PRINCIPLE OF THE METHOD

1.1 Solutions with zinc compounds are nebulised into an argon plasma, where all components are vaporised. The ions produced are entrained in the plasma gas and introduced into a mass spectrometer, sorted according to their mass-to-charge ratios, and quantified with a channel electron multiplier. Zinc is determined at mass 66 or 68 amu.

1.2 This determination may be carried out on digest 2.3 (microwave digestion with $HNO_3$ - $H_2O_2$ - HF), digest 2.4 (digestion with $HNO_3$ - $H_2O_2$ - HF) and digest 2.7 (digestion by dry-ashing followed by treatment with HF).

## 2. RANGE AND DETECTION LIMIT

2.1 This procedure yields a standard curve that is linear up to at least 50 μg/L Zn.

2.2 The detection limit is approximately 1.5 μg/L in the digest. The determination limit is approximately 4.5 μg/L (0.5 respectively 1.5 mg/kg in the dried plant material).

## 3. INTERFERENCES

3.1 Interferences can be expected from $^{50}Ti^{16}O$, $^{34}S^{16}O_2$, $^{132}Ba^{++}$ due to mass overlap at $^{66}Zn$. $^{136}Ba^{++}$, $^{136}Ce^{++}$, $^{40}Ar^{14}N_2$, $^{54}Fe^{14}N$ show mass overlap at $^{68}Zn$.

## 4. PRECISION AND ACCURACY

4.1 The reproducibility of determinations by this procedure should give, at reasonable control and thorough sample preparation, a coefficient of variation within 10 %.

## 5. APPARATUS

5.1 Inductively coupled plasma mass spectrometer.

## 6. REAGENTS

6.1A Stock Solution, Zn concentration 1000 mg/L - Merck nr 1.02670.

6.1B Stock Solution, Zn concentration 1000 mg/L - Dissolve 4.398 g zinc sulphate heptahydrate, $ZnSO_4.7H_2O$, in some ultra pure water in a 1000-mL volumetric flask. Make up to 1000 mL with ultra pure water.

6.2 Standard Solution, Zn concentration 1 mg/L - Pipette 1 mL concentrated nitric acid (65 % s.p.) in a 1000-mL polythene volumetric flask which already contains about

500 mL ultra pure water. Add 1.00 mL stock solution (6.1A or 6.1B) and make up to volume with ultra pure water.

Remark:

1. $ZnSO_4.7H_2O$ may lose crystal water on standing. The reagent should be standardised by titration with EDTA at pH=10 with Eriochrome Black T as an indicator.

# 7. CALIBRATION AND STANDARDS

7.1 Standard Series - pipette 0 – 1.00 – 2.00 – 5.00 mL of the standard solution (6.2) in about 40 mL ultra pure water in a 100-mL polythene volumetric flasks. Add either 1.0 mL concentrated nitric acid (65 %) (digestion 2.3), 0.3 mL concentrated nitric acid (65 %) (digestion 2.4) or 0.1 mL concentrated hydrochloric acid (36 %) (digestion 2.7). Let cool down and make up to the mark with ultra pure water. This standard series has Zn concentrations of 0 –10 – 20 – 50 µg/L.

7.2 Calibration Curve - The counts per second are plot versus µg/L zinc in the standard series.

Remark:

2. A mix standard series can be used for simultaneous measurements with ICP-MS (Al, As, Cd, Co, Cr, Cu, Mn, Ni, Pb, Sb, Sn, V and Zn).

# 8. PROCEDURE

8.1 Measurement - Dilute the blanks and the sample digests 1 + 9 (v/v) with ultra pure water and mix. Measure in the standard series, the blanks and the sample digests the Zn concentration with the ICP-MS at a mass of 66 or 68 amu. Make use of corrections if necessary.

Remarks:

3. A data management system and system controller are used. In this way the measurements are checked continuously and the data output is in concentration units in the digests.
4. Due to the linearity of the ICP-MS it is possible to calibrate the ICP-MS only with the highest and zero standard of the standard series. As a check all standards can be measured as samples.

# 9. CALCULATION

9.1 The total zinc content in the dried plant material, expressed in mg/kg Zn, is calculated by:

$0.001*(a - b) * V / w$

in which:

a is the concentration of zinc in the sample digest, in µg/L;

b is the concentration of zinc in the blank digest, in µg/L;

V is the total volume of digest at the end of the digestion procedure, in mL;

w is the weight of plant material sample, in g.

MIX
Papier aus verantwortungsvollen Quellen
Paper from responsible sources
FSC® C105338
FSC
www.fsc.org

If you have any concerns about our products,
you can contact us on
**ProductSafety@springernature.com**

In case Publisher is established outside the EU,
the EU authorized representative is:
**Springer Nature Customer Service Center GmbH**
**Europaplatz 3, 69115 Heidelberg, Germany**

Printed by Libri Plureos GmbH
in Hamburg, Germany